# Lecture Notes in Computer Science 6260

Commenced Publication in 1973
Founding and Former Series Editors:
Gerhard Goos, Juris Hartmanis, and Jan van Leeuwen

## Editorial Board

David Hutchison
*Lancaster University, UK*

Takeo Kanade
*Carnegie Mellon University, Pittsburgh, PA, USA*

Josef Kittler
*University of Surrey, Guildford, UK*

Jon M. Kleinberg
*Cornell University, Ithaca, NY, USA*

Friedemann Mattern
*ETH Zurich, Switzerland*

John C. Mitchell
*Stanford University, CA, USA*

Moni Naor
*Weizmann Institute of Science, Rehovot, Israel*

Oscar Nierstrasz
*University of Bern, Switzerland*

C. Pandu Rangan
*Indian Institute of Technology, Madras, India*

Bernhard Steffen
*TU Dortmund University, Germany*

Madhu Sudan
*Microsoft Research, Cambridge, MA, USA*

Demetri Terzopoulos
*University of California, Los Angeles, CA, USA*

Doug Tygar
*University of California, Berkeley, CA, USA*

Moshe Y. Vardi
*Rice University, Houston, TX, USA*

Gerhard Weikum
*Max Planck Institute for Informatics, Saarbruecken, Germany*

W0192937

Marina L. Gavrilova   C.J. Kenneth Tan (Eds.)

# Transactions on Computational Science VIII

 Springer

Editors-in-Chief

Marina L. Gavrilova
University of Calgary, Department of Computer Science
2500 University Drive N.W., Calgary, AB, T2N 1N4, Canada
E-mail: mgavrilo@ucalgary.ca

C.J. Kenneth Tan
Exascala Ltd.
Unit 9, 97 Rickman Drive, Birmingham B15 2AL, UK
E-mail: cjtan@exascala.com

Library of Congress Control Number: 2010935450

CR Subject Classification (1998): F.2, F.1, I.4, J.3, I.5, G.2

| ISSN | 0302-9743 (Lecture Notes in Computer Science) |
| ISSN | 1866-4733 (Transaction on Computational Science) |
| ISBN-10 | 3-642-16235-5 Springer Berlin Heidelberg New York |
| ISBN-13 | 978-3-642-16235-0 Springer Berlin Heidelberg New York |

springer.com

© Springer-Verlag Berlin Heidelberg 2010
Printed in Germany

Typesetting: Camera-ready by author, data conversion by Scientific Publishing Services, Chennai, India
Printed on acid-free paper        06/3180

# LNCS Transactions on Computational Science

Computational science, an emerging and increasingly vital field, is now widely recognized as an integral part of scientific and technical investigations, affecting researchers and practitioners in areas ranging from aerospace and automotive research to biochemistry, electronics, geosciences, mathematics, and physics. Computer systems research and the exploitation of applied research naturally complement each other. The increased complexity of many challenges in computational science demands the use of supercomputing, parallel processing, sophisticated algorithms, and advanced system software and architecture. It is therefore invaluable to have input by systems research experts in applied computational science research.

*Transactions on Computational Science* focuses on original high-quality research in the realm of computational science in parallel and distributed environments, also encompassing the underlying theoretical foundations and the applications of large-scale computation. The journal offers practitioners and researchers the opportunity to share computational techniques and solutions in this area, to identify new issues, and to shape future directions for research, and it enables industrial users to apply leading-edge, large-scale, high-performance computational methods.

In addition to addressing various research and application issues, the journal aims to present material that is validated – crucial to the application and advancement of the research conducted in academic and industrial settings. In this spirit, the journal focuses on publications that present results and computational techniques that are verifiable.

## Scope

The scope of the journal includes, but is not limited to, the following computational methods and applications:

- Aeronautics and Aerospace
- Astrophysics
- Bioinformatics
- Climate and Weather Modeling
- Communication and Data Networks
- Compilers and Operating Systems
- Computer Graphics
- Computational Biology
- Computational Chemistry
- Computational Finance and Econometrics
- Computational Fluid Dynamics
- Computational Geometry

- Computational Number Theory
- Computational Physics
- Data Storage and Information Retrieval
- Data Mining and Data Warehousing
- Grid Computing
- Hardware/Software Co-design
- High-Energy Physics
- High-Performance Computing
- Numerical and Scientific Computing
- Parallel and Distributed Computing
- Reconfigurable Hardware
- Scientific Visualization
- Supercomputing
- System-on-Chip Design and Engineering

# Editorial

The Transactions on Computational Science journal is part of the Springer series *Lecture Notes in Computer Science*, and is devoted to the gamut of computational science issues, from theoretical aspects to application-dependent studies and the validation of emerging technologies.

The journal focuses on original high-quality research in the realm of computational science in parallel and distributed environments, encompassing the facilitating theoretical foundations and the applications of large-scale computations and massive data processing. Practitioners and researchers share computational techniques and solutions in the area, identify new issues, and shape future directions for research, as well as enable industrial users to apply the techniques presented.

The current issue is comprised of two parts. Part 1 focuses on adaptive evolutionary computation, and was prepared by Guest Editors Nadia Nedjah, Abdelhamid Bouchachia and Luiza de Macedo Mourelle. The focus of Part 2 is on computational methods for model visualization and analysis.

Part I consists of five manuscripts, each addressing a specific computational problem utilizing adaptive methodologies. It also contains a comprehensive review of the current advancements in the area prepared by the Special Issue Guest Editors.

Part II continues the theme. It is comprised of six manuscripts that take an in-depth look at selected computational science research in the areas of geometric computing, Euclidean distance transform, distributed systems, segmentation, visualization of monotone data and data interpolation. Each paper provides a rigorous analysis or a detailed experiment to amplify the impact of the contribution.

In conclusion, we would like to extend our sincere appreciation to Special Issue Guest Editors Nadia Nedjah, Abdelhamid Bouchachia and Luiza de Macedo Mourelle for their efficiency and diligence, all authors for submitting their papers, and all Associate Editors and referees for their valuable work. We would also like to express our gratitude to the LNCS editorial staff of Springer, in particular Alfred Hofmann, Ursula Barth and Anna Kramer, who supported us at every stage of the project.

It is our hope that the fine collection of papers presented in this journal issue will be a valuable resource for Transactions on Computational Science readers and will stimulate further research into the vibrant area of computational science applications.

June 2010

Marina L. Gavrilova
C.J. Kenneth Tan

# Adaptive Evolutionary Computation

## Special Issue Guest Editors' Preface

Adaptation plays a central role in dynamically changing systems. It is about the ability of a system to "responsively" self-adjust in response to change in the surrounding environment. Like living creatures that have evolved over millions of years developing ecological systems due to their self-adaptation and fitness capacity to the dynamic environment, systems undergo similar cycles to improve or at least to not weaken their performance when internal or external changes take place. Internal change bears on the physical structure of the system (the building blocks: hardware and/or software components). External change originates from the environment due to reciprocal action and interaction. These two classes of change shed light on the research avenues towards smart adaptive systems.

The state of the art draws the picture of challenges that such systems need to face before they become reality. A sustainable effort is necessary to develop intelligent hardware on one level and concepts and algorithms on the other level. The former level concerns various analog and digital accommodations encompassing self-healing, self-testing, reconfiguration and many other aspects of system development and maintenance. The latter level is concerned with developing algorithms, concepts and techniques that can rely on metaphors of nature and that are inspired from biological and cognitive plausibility.

To face the different types of change, systems must self-adapt their structure and self-adjust their controlling parameters over time as changes are sensed. A fundamental issue is the notion of "self", which refers to the capability of the systems to act and react on their own. It covers all stages of the systems working and maintenance cycle starting from online self-monitoring to self-growing and self-organizing. Relying on the two-fold plausibility, which is the basis for many computational models, neural networks can be encountered in various real-world dynamical and non-stationary systems that require continuous update over time. There exist many neural models that are theoretically based on incremental (i.e., online, sequential) learning, addressing in particular the notions of self-growing and self-organizing. However, their strength in practical situations that involve online adaptation is not as efficient as desirable.

Part I of this journal aims to present the latest advances of adaptive models for evolutionary computation and their application in various dynamic environments. The special issue is intended for a wide audience including scientists as well as mathematicians, physicists, engineers, computer scientists, biologists, economists and social scientists. This special section covers various topics of evolutionary computation related to self-organization, self-monitoring and self-growing concepts. It also aims at presenting a coherent view of these issues and a thorough discussion about the future research avenues. The main contribution of each of the five papers included is briefly introduced in the following.

In the first paper of the section, entitled "Environmental Modeling and Identification Based on Changes in Sensory Information", the authors present an environmental modeling method based on state representation, which represents a change in sensory information. The model presented enables the mobile robot to identify which environment it is in. The results of experiments on a real mobile robot with only low-sensitivity infrared sensors show the effectiveness of the method.

In the second paper, entitled "Polymorphic Particle Swarm Optimization", the author proposes a new concept called polymorphic particle swarm optimization, which generalizes the standard update rule by a polymorphic update rule. This polymorphic update rule is an adaptive update rule changing symbols based on accumulative histograms and roulette-wheel sampling. The proposed variant is applied to typical benchmark functions and in most cases it outperforms the other PSO existing variants.

In the third paper, entitled "C-Strategy: A Dynamic Adaptive Strategy for the CLONALG Algorithm", the authors propose a new parameter control strategy for the immune algorithm CLONALG. The approach presented is based on the concepts behind reinforcement learning. The approach provides an efficient and low cost adaptive technique for parameter control. The results obtained are very encouraging.

In the fourth paper, entitled "A Comparison of Genotype Representations to Acquire Stock Trading Strategy Using Genetic Algorithms", the authors compare some genotype coding methods of technical indicators and their parameters to acquire a stock trading strategy using genetic algorithms. They show that these conventional representations are not so effective for the GA search. Thereafter, they propose a new genotype coding methods, namely the allele-based indirect representation and show the superiority of the proposed individual representation.

In the fifth paper, entitled "Automatic Adaptive Modeling of Fuzzy Systems Using Particle Swarm Optimization", the authors show how to yield adaptive fuzzy models by applying a particle swarm-based optimization method. The results of this automatic modeling are given for three complex three-dimensional functions and prove the effectiveness of the proposed method.

The Guest Editors are very grateful to the authors of this special section and to the reviewers for their tremendous service in critically reviewing the submitted papers. The editors would also like to thank Prof. Marina Gavrilova, the Editor-in-Chief of Transactions on Computational Science, Springer-Verlag, for the editorial assistance and excellent cooperative collaboration in producing this scientific work. We hope that the reader will share our excitement about the papers on adaptive evolutionary computation and will find them useful.

June 2010

Nadia Nedjah
Abdelhamid Bouchachia
Luiza de Macedo Mourelle

# LNCS Transactions on Computational Science – Editorial Board

# Table of Contents

## Part I: Adaptive Evolutionary Computation

Environmental Modeling and Identification Based on Changes in
Sensory Information .............................................. 3
    *Manabu Gouko and Koji Ito*

Polymorphic Particle Swarm Optimization ......................... 20
    *Christian Veenhuis*

C-Strategy: A Dynamic Adaptive Strategy for the CLONALG
Algorithm ........................................................ 41
    *María Cristina Riff, Elizabeth Montero, and Bertrand Neveu*

A Comparison of Genotype Representations to Acquire Stock Trading
Strategy Using Genetic Algorithms ................................ 56
    *Kazuhiro Matsui and Haruo Sato*

Automatic Adaptive Modeling of Fuzzy Systems Using Particle Swarm
Optimization ..................................................... 71
    *Sergio Oliveira Costa Jr., Nadia Nedjah, and
    Luiza de Macedo Mourelle*

## Part II: Computational Methods for Model Visualization and Analysis

Computational Algorithm for Some Problems with Variable
Geometrical Structure ............................................ 87
    *N. Bessonov and V. Volpert*

In-Place Linear-Time Algorithms for Euclidean Distance Transform .... 103
    *Tetsuo Asano and Hiroshi Tanaka*

A Foundation of Demand-Side Resource Management in Distributed
Systems .......................................................... 114
    *Shrisha Rao*

Modified Bias Field Fuzzy C-Means for Effective Segmentation of Brain
MRI .............................................................. 127
    *S.R. Kannan, S. Ramathilagam, and R. Pandiyarajan*

Visualization of Monotone Data by Rational Bi-cubic Interpolation . . . . .    146
    *Malik Zawwar Hussain, Maria Hussain, and Muhammad Sarfraz*

$C^1$ Monotone Scattered Data Interpolation . . . . . . . . . . . . . . . . . . . . . . . . .    156
    *Malik Zawwar Hussain and Maria Hussain*

**Author Index** . . . . . . . . . . . . . . . . . . . . . . . . . . . . . . . . . . . . . . . . . . . . . . . .    167

# Part I

## Adaptive Evolutionary Computation

# Environmental Modeling and Identification Based on Changes in Sensory Information

Manabu Gouko[1] and Koji Ito[2]

[1] Dept. of Electrical and Electronic Engineering,
College of Engineering, Nihon University,
1 Nakagawara, Tokusada, Tamuramachi,
Koriyama, Fukushima, 963-8642, Japan
gouko@ee.ce.nihon-u.ac.jp
[2] Ritsumeikan University, Research organization of science and engineering
1-1-1 Noji Higashi, Kusatsu, Shiga 525-8577, Japan
koji-ito@fc.ritsumei.ac.jp

**Abstract.** Adaptability to various environments is needed for a robot that supports our lives. Environmental identification is important for a mobile robot that works in multiple environments (e.g., different rooms). We present an environmental modeling method based on state representation, which represents a change in sensory information. Our model enables the mobile robot to identify which environment it is in. The results of experiments on a real mobile robot with only low-sensitivity infrared sensors show the effectiveness of our method, and a comparison between our method and a conventional one shows that ours has higher performance.

## 1 Introduction

In research on autonomous mobile robots, many studies on localization and navigation tasks have been conducted. These tasks are generally performed using prior knowledge (e.g., a map of the experiment) based on visual and position data of its environment [1–4]. For example, a robot that works in several environments, such as different rooms, needs knowledge about each environment. If the robot can recognize that it is in a known environment, it can use the map of the surroundings. On the other hand, if the robot recognizes that it is in an unknown environment, it will try to build a new map of the environment. Thus, it is important for a robot to identify which environment it is currently in so that it can work efficiently. Autonomous mobile robots should be able to identify their environments by themselves.

An easy method of identifying multiple environments is to set landmarks in each environment. The robot can identify which environment it is in by comparing the landmarks; however, this requires space and time for installation. Another method is to use environmental modeling and comparison (Figure 1). The robot

M.L. Gavrilova et al. (Eds.): Trans. on Comput. Sci. VIII, LNCS 6260, pp. 3–19, 2010.
© Springer-Verlag Berlin Heidelberg 2010

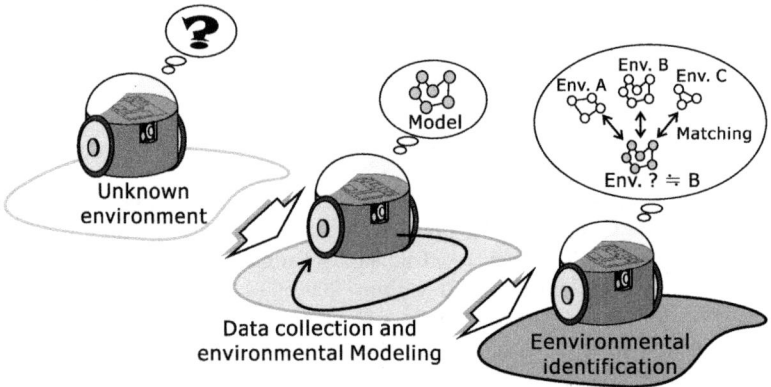

**Fig. 1.** Overview of the environmental identification

makes a model of its current environment and compares it against many stored environmental models. For highly accurate recognition, this method requires a model that can represent the characteristics of a real environment. Moreover, the model must be easier to make than a new map.

Research on environmental modeling using stochastic models is well known. Models built using such an approach are applicable not only to environmental discernment but also to localization and navigation. However, building a stochastic model requires a lot of data, which is collected by making the robot move about in the environment. Therefore, it is difficult to use this approach to enable a real robot to make multiple environmental models.

The characteristics (e.g., size and layout) of each environment appear as the form of the environment if the environment is a closed region surrounded by walls. On the assumption that the environment is a closed region, several studies on environmental modeling and identification have been done[1, 5, 6].

Mataric proposed a modeling method for integrating two sets of information from sonar sensors and from a compass[1]. She defined several features about the form of the walls using sensory information and then made a model using those features called landmarks. The model was represented by a graph with nodes corresponding to the landmarks. In that study, the effectiveness of the method was evaluated by using the model for path planning. The landmark-based model is robust against sensor noise, but, in this method, a designer must make the raw sensory data correspond to the landmarks prior to the modeling.

Nehmzow et al. proposed an environmental modeling method using sensory data acquired by making the mobile robot perform a wall-following movement[5]. They used information about the kind the corner (e.g., convex or concave) and the distances between two adjoining corners as environmental characteristics. They input these features into a neural network as input vectors, and the learned network could then identify the forms of the environment. However, this work treated only simple environments with right-angled corners and straight walls.

Yamada and Murota proposed a modeling method and tested multiple environment recognition by a real mobile robot[6]. In that method, the robot initially performs a wall-following movement in its environment under the control of if-then rules. Then, the rule sequence observed during wall-following is used for modeling. If environmental models are made using if-then rules instead of using sensor data directly, the acquired models are robust against sensory noise. They confirmed the effectiveness of their method through an experiment with a real mobile robot that had only low-sensitivity infrared sensors. By using the models built by this method, they demonstrated that it was possible to distinguish seven environments having different shapes. However, the identification performance depended on the rules, which were designed before the modeling. This means that the performance of their method is unreliable if suitable rules cannot be determined.

Many of these conventional studies have created an environmental model by making raw sensor data correspond to state representations (e.g., if-then rules, landmarks, and corners). However, the performances of these methods are unreliable if suitable state representations cannot be determined.

In this paper, we present a modeling method based on state representation, which represents a change in sensory information. There have been several studies on methods of using changes in sensory information for robot state representation [7–9]. Duchon et al. used optical flow as a state representation and used it to control a mobile robot [7]. Takahashi et al. proposed a state representation based on changes in sensory output for reinforcement learning.

Nakamura et al. proposed an environmental modeling method based on changes in sensory information [9]. They used the change observed by the moving robot for the model. However, the purpose of that study was to make a model of the whole environment for use in robot action generation. The method needs a lot of data and statistical processing. Moreover, the parameters needed for statistical processing depend strongly on the environment. Such an approach is not suitable for modeling and identifying multiple environments, which is the topic treated in this paper.

In this paper, we identify an environment by using a model based on changes in sensory information. The model is built from changes in sensory data observed by the mobile robot during wall-following movements. Our method is targeted at mobile robots that have only low-sensitivity short-range sensors because there are cost advantages if such a simple robot can identify its environment. In experiments, we confirmed the effectiveness of our method by applying it to a mobile robot having only low-sensitivity infrared sensors. We found that highly accurate identification could be achieved by using our model.

This rest of this paper is organized as follows. Section 2 describes the method of modeling the environment and the method of identifying multiple environments by using the model. Section 3 presents experimental results obtained with a real mobile robot. Section 4 discusses the experimental results. Section 5 concludes with a summary and mentions future work.

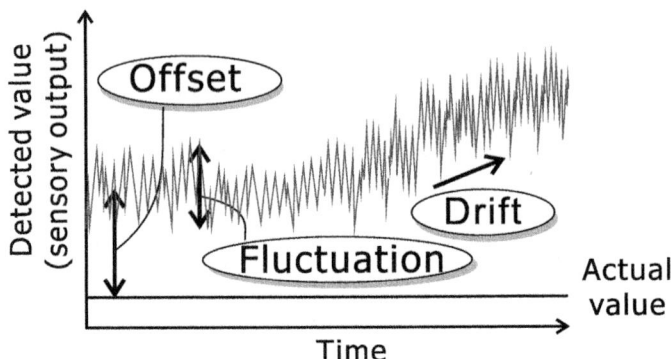

**Fig. 2.** Sensory noise

## 2    Proposed Method

This section describes the method of environmental modeling and the method of identifying multiple environments by using the model. Proposing a method for even a simple robot is valuable from the engineering viewpoint. Although the method described below is for a mobile robot with only low-sensitivity infrared sensors, it should be robust against changes in lighting conditions.

### 2.1    Sensory Noise

First, we discuss general sensory noise. Types of sensor noise are shown in Figure 2. Fluctuation and drift represent changes in comparatively quick and late sensory output, respectively.

Offset means the difference between an actual value and a detected value. Since sensory output generally includes such noise, frequent calibration is necessary. From the viewpoint of environment modeling, these noises reduce the similarity of models built for the same environment, which makes highly accurate identification difficult. Moreover, variations in the robot's movement also makes highly accurate identification difficult. Even if the same motor command is output from a controller, an actual robot's action will differ because of friction or mechanism factors. Therefore, even when the robot moves in response to the same-shaped wall, the obtained sensory patterns will differ (figure 3). Below, we propose an environmental modeling method that is robust against these sensory noises and the effects of movement variations.

### 2.2    State Representation

We assume that the robot has multiple distance sensors, and we define the pattern $s_t = (s_{t,1}, \cdots, s_{t,i}, \cdots, s_{t,I})$ as the sensory pattern observed at time $t$, where $s_{t,i}$ represents the output of sensor number $i$ at time $t$. In this paper, for

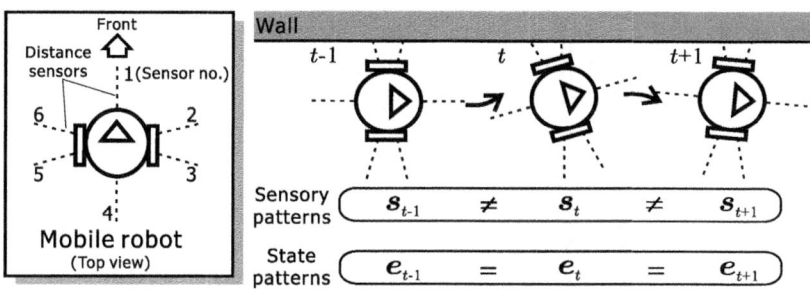

**Fig. 3.** Dispersion of movement during wall-following

simplicity, we assume that all of the sensors have the same specifications. The sensor pattern series observed by the robot as it moves round the environment along a wall are used for modeling. Such a movement in which the robot goes round the environment along a wall is called a trial. And the observed sequence is defined as $s_0, \cdots, s_t, \cdots, s_T$. In our method, the change in sensory patterns between times $t$ and $t+1$ is calculated in order to remove the effect of sensory offset. The sensory pattern change is defined as $\Delta s_{t,i} = s_{t,i} - s_{t-1,i}$. Here, $\Delta s_{t,i}$ is not affected by sensory offset because it is a relative amount. Furthermore, if the rate of change of drift is very slow, $\Delta s_{t,i}$ will not be influenced by drift, either.

Next, we consider the effect of variations in the robot's behavior. Motion variation produces different values of $\Delta s_{t,i}$ even if the robot moves along the same-shaped wall. $\Delta s_{t,i}$ consists of two components of change.

$$\Delta s_{t,i} = \Delta s_{t,i}^f + \Delta s_{t,i}^m, \tag{1}$$

where $\Delta s_{t,i}^f$ is the change resulting from sensory fluctuation and $\Delta s_{t,i}^m$ is the change resulting from robot movement. $\Delta s_{t,i}^f$ is non-zero even if the robot does not move. $\Delta s_{t,i}^m$ arises from the change in actual distance between the robot and the wall caused by the robot's movement. For example, in figure 3, $\Delta s_{t,i}$ for sensors 1 to 4 consists of only $\Delta s_{t,i}^f$. Because the wall-to-robot distance in the sensing area changes with the movement, $\Delta s_{t,i}$ for sensors 5 and 6 consists of $\Delta s_{t,i}^f$ and $\Delta s_{t,i}^m$.

We assume that $|\Delta s_{t,i}^f|$ is very small compared with $|\Delta s_{t,i}^m|$. In the case of figure 3, this assumption means that $\Delta s_{t,i}$ for sensors 5 and 6 is always larger than $\Delta s_{t,i}$ for the other sensors. In our method, such a relationship between the $|\Delta s_{t,i}|$ values of sensors is used as a pattern that represents the shape of walls. A pattern $e_t = (e_{t,1}, \cdots, e_{t,i}, \cdots, e_{t,I})$ is defined as

$$e_{t,i} = \begin{cases} 1 & \text{if } |\Delta s_{t,i}| > \eta \\ 0 & \text{otherwise} \end{cases}$$

$$(t = 1, 2, \cdots, T_e, \quad i = 1, 2, \cdots, I), \tag{2}$$

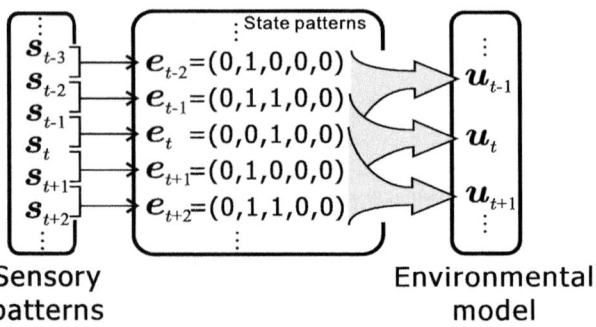

**Fig. 4.** Proposed method ($N = 1, I = 5$)

where $\eta$ is a threshold and $T_e$ is larger than $T$. Even if the sensory pattern varies, the same pattern $\boldsymbol{e}_t$ will be obtained if we set $\eta$ using $|\Delta s_i^m| > \eta > |\Delta s_i^f|$ (figure 3). In this paper, we call $\boldsymbol{e}_t$ the state pattern at time $t$.

In Equation (2), if $t$ is larger than $T$, we set $\boldsymbol{e}_t$ to a zero vector. If $T_e$ is set large enough, state pattern sequences of the same length are acquired from environments of different sizes. The proposed state pattern uses information about the existence of a wall within sensor range. It uses the change in sensory output to detect the wall. There is also a way to detect the wall from sensory output directly. The wall can be detected if the robot judges that a wall exists when the sensory output is larger than a threshold. However, if the sensor offset differs for every trial, it is difficult to identify the existence of a wall accurately. This method needs to tune the threshold frequently. On the other hand, our method can be applied to data that includes a different offset for every trial. Additionally, the use of a different threshold $\eta$ for each sensor makes our method applicable to robots with sensors having different specifications.

## 2.3   Environmental Modeling and Identification

The environmental modeling method is described below. First, the sequence of the state pattern is made from the sensory pattern sequence. Then, a vector $\boldsymbol{u}_k$ is calculated as

$$\boldsymbol{u}_k = \frac{1}{2N+1} \sum_{j=k}^{k+2N} \boldsymbol{e}_j$$
$$(k = 1, 2, \cdots, T_e - 2N), \tag{3}$$

where $\boldsymbol{u}_k$ is the average of $\boldsymbol{e}_t$. We defined the sequence $\boldsymbol{u}_1, \cdots, \boldsymbol{u}_k, \cdots, \boldsymbol{u}_{T_e-2N}$ as the model of the environment. The procedure for the modeling ($N = 1, I = 5$) is shown in Figure 4.

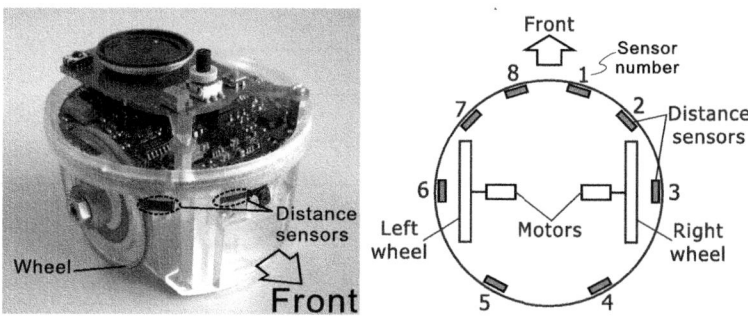

**Fig. 5.** Mobile robot (e-puck)

Environment identification is achieved by model comparison. Our method identifies the environment by calculating the error between the model of the current environment (test model) and the reference models.

$$E_l = \sum_{k=1}^{T_e-2N} (\boldsymbol{u}_k^l - \boldsymbol{u}_k^{test})^2 \quad (l = 1, 2, \cdots, L), \qquad (4)$$

where $\boldsymbol{u}_k^l$ and $\boldsymbol{u}_k^{test}$ are elements in the sequence of test and reference models, respectively, and $E_l$ is the error between the test model and the reference model of environment $l$. The recognition result $\hat{l}$ is given as $\hat{l} = \text{argmin}_l E_l$.

## 3  Experimental Results and Discussion

To check the effectiveness of our method, we performed experiments using a real mobile robot.

### 3.1  Experimental Setup

The mobile robot used in the experiments is shown in Figure 5 [10]. It is called e-puck. It is cylindrical with a diameter of 70 mm and a height of 40 mm and it has eight distance sensors (infrared sensors) and two wheels. The sensor detection ranges are 40 mm. All the sensor readings were normalized to the range [0,1], where 1 indicates that the robot was very close to an object and 0 indicates that there was no object within the sensor's detection range. The sensor pattern $\boldsymbol{s}_t$ was denoted by a vector of order 8 ($I = 8$). Its eight elements correspond to the output values of sensors 1 to 8, respectively. The first experiment used the seven simple environments shown in Figure 6. These environments, made using polystyrene boards, consisted of straight walls and right-angled corners (figure 7). We compared the performance of our method (proposed) with AEM, which is one of the environmental identification methods proposed by Yamada and Murota [6].

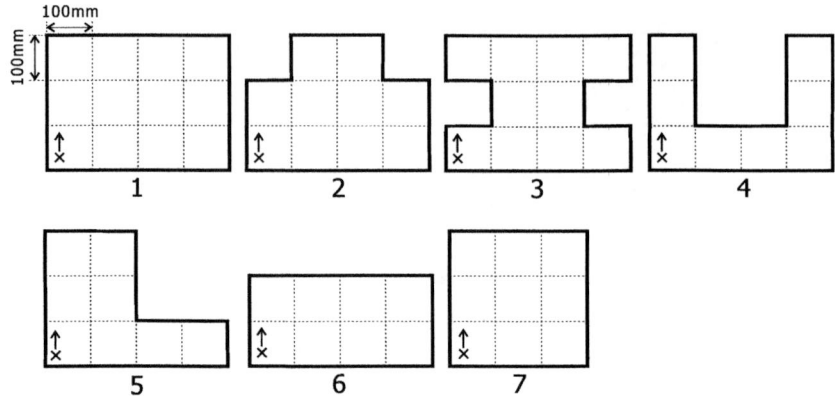

**Fig. 6.** Environments (1–7)

AEM uses a mobile robot controlled by if-then rules, so identification is achieved using the model it makes from the sequence of rules used. For environments similar to our seven in Figure 6, they reported that their method could identify all of them accurately. In our experiments, we used a mobile robot controlled by if-then rules. The sequence of sensor patterns was obtained when the robot moved round inside the environments. The if-then rules used are shown below.

*Rule A* (turning in a concave corner)
    **If** $s_8 > 0.3$.
    **Then** turn 16 degrees clockwiseD


*Rule B* (turning in a convex corner)
    **If** $s_6 + s_7 < 0.2$.
    **Then** turn 16 degrees counterclockwise and go straight for 10 mmD


*Rule C* (following a wall)
    **If** $(s_6 + s_7)/2 > 0.3$.
    **Then** turn 2 degrees clockwise and go straight for 3.3 mm.


*Rule D* (following a wall)
    **If** $(s_6 + s_7)/2 \leq 0.3$.
    **Then** turning 2 degrees counterclockwise and go straight for 3.3 mm.

In each rule, the "if" part represents the condition described by sensory output, and the "then" part represents the movement generated by the robot. At time $t$, the robot selects a rule from sensory pattern $s_t$, and a movement is generated according to this rule.

**Fig. 7.** Experimental environment

The robot observed the sensory patterns and generated a behavior at every time step (0.3 s). Even when it used the same rule, its action varied as a result of friction. The robot moved according to the selected rule for 0.3 s, except when the selected rule was $B$, in which case the robot rotated and went straight for 0.3 s. Rules $A$ and $B$ were used mainly at corners, and rules $C$ and $D$ were used mainly when the robot moved along in accordance with straight walls.

In the experiments, we made the robot go around the environments at first and we obtained six sequences (trials) from each environment. 42 models were made for each sequence (7 environments×6 trials). One model was randomly chosen for every environment and treated as a reference model. The other models were used as test models. The recognition rate was defined as the ratio of the number of tested models that were recognized correctly to the total number of tested models as follows.

Recognition rate (%)

$$= \frac{\text{Number of correctly recognized tested models}}{\text{Total number of tested models}} \times 100 \qquad (5)$$

The recognition rate showed hereinafter is the average of all the rates calculated by replacing the reference model.

We assumed that there were different sensor offsets for every trial, so we added noise to the sensory patterns of each sequence. The noise level was set between 0 and 0.1 on the basis of the actual sensor offset. The maximum sequence length was 400, so we used $T_e = 450$. We also investigated the recognition rate for two other state representations: $s_t^{WALL}$ and $s^{SENS}$.

$s_t^{WALL}$ was obtained by using Equation (2) with $|\Delta s_{t,i}|$ and $\eta$ replaced by $s_{t,i}$ and $\mu$, respectively. Here, $s_t^{WALL}$ represents the existence of a wall determined using threshold $\mu$. The sensory pattern $s_t$ was used as $s^{SENS}$. The methods using $s_t^{WALL}$ and $s_t^{SENS}$ are denoted "Wall" and "Sensor".

**Table 1.** Recognition rates for seven environments

| Method | Maximum recognition rate |
|:---:|:---:|
| **Proposed** | **100%** |
| AEM | 100% |
| Wall | 97.6% |
| Sensor | 93.3% |

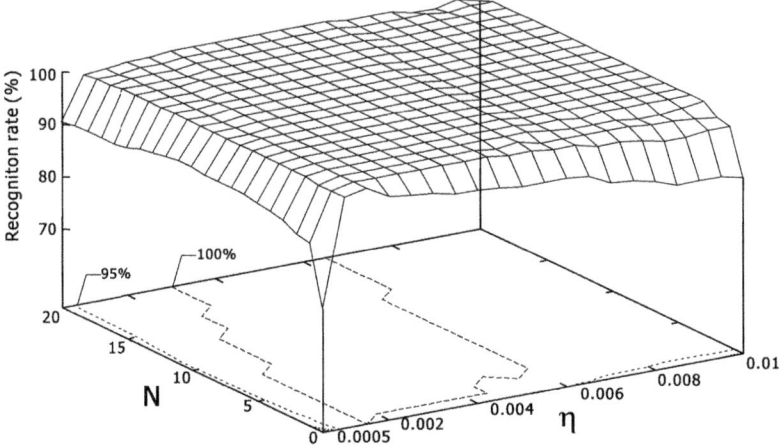

**Fig. 8.** Recognition rates for proposed method using several parameter sets (7 environments)

## 3.2   Identification of 7 Environments

In this experiment, we changed parameters $N$, $\eta$, and $\mu$ and investigated the recognition rate. The maximum rates obtained are listed in Table 1. Our method and AEM achieved 100%. The recognition rates for our method and for Wall, which were acquired by changing the parameters, are shown in Figures 8 and 9. They indicate that the recognition rate of our method does not depend on the parameters as much. The maximum rate for Wall was obtained with $\mu = 0.25$ and $N = 7$, and the maximum for Sensor was obtained with $N = 9$.

## 3.3   Identification of 13 Environments

In this experiment, we investigated the recognition rate for 13 environments. We added the six new environments shown in Figure 10 to the seven environments used before. New environments 8, 9, and 10 were similar to environments 1, 6, and 8; they differed only in the corner angle. Environments 11, 12, and 13 were also similar to environments 1, 6, and 8, but each one had a curved wall.

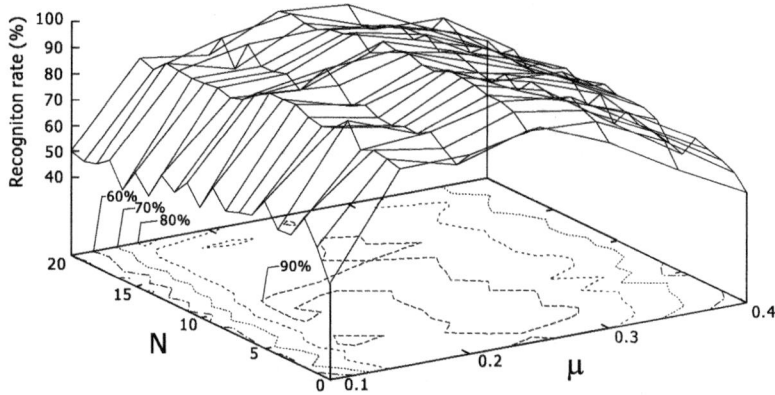

**Fig. 9.** Recognition rates for Wall using several parameter sets (7 environments)

**Table 2.** Recognition rates for 13 environments

| Method | Maximum recognition rate | Recognition rate using same parameters for 7 environments |
|---|---|---|
| **Proposed** | **99.0%** | **95.6%** |
| AEM | 77.9% | - |
| Wall | 82.3% | 76.7% |
| Sensor | 84.6% | 83.3% |

The wall curvature in environments 11, 12, and 13 were 1.2, 4.2, and 2.9 [1/m], respectively.

The maximum rates are listed in Table 2. The maximum rate for our method was 99%, obtained with $\eta = 0.002 - 0.0025$ and $N = 11 - 13$. Thus, our method was able to identify the new environments. On the other hand, the maximum rate for Wall ($\mu = 0.25$ and $N = 13$) and Sensor ($N = 13$) decreased, which shows that these methods could not identify the newly added similar environments by analyzing the error in identification. The recognition rate of AEM also decreased. Almost all the mistakes occurred in similar environments. When rules $C$ and $D$ were used, AEM considered the straight wall and modeled it. Therefore, AEM could not distinguish between the straight and curved walls. Moreover, it had trouble identifying the difference in corner angle. Although AEM can identify the corner angle from the number of times the rule is used, corner angle recognition using these rules was difficult. We conclude that deciding the rules in advance was the cause of this decline in recognition rate.

We investigated the recognition rate using the same parameters as used for the maximum rate in the seven-environment identification (Table 2). Our method could distinguish them better than the other methods. This means that our method's discernment performance is robust against changes in environment.

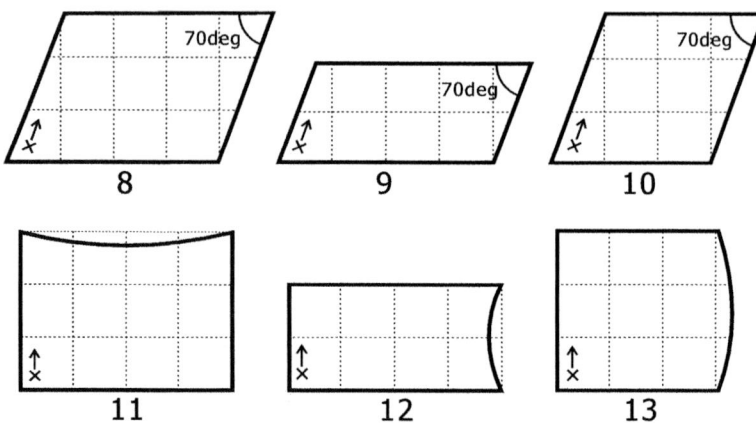

**Fig. 10.** Environments (8–13)

The recognition rates for our method and for Wall, which were acquired by changing the parameters, are shown in Figures 11 and 12.

Parameter $\eta$ of the maximum rate decreased, as shown in Figure 11. Our method can express a state finely when the threshold $\eta$ is small. However, a small $\eta$ causes a decrease in recognition rate as a result of the influence of fluctuations in sensory output. It is effective to measure $s_t$ in advance and use the results as a criterion for determining the threshold $\eta$. We infer that an appropriate threshold can be determined by increasing the parameter gradually from a small value. The determination of an appropriate value for $\eta$ will be studied in future work.

To check whether the state expression reflected the environmental form, we performed a cluster analysis of state pattern $e_t$. In this analysis, the state patterns obtained from environments 7, 10, and 13 were used. We applied the $k$-means clustering algorithm to the state patterns obtained in each environment. Parameter $k$ was set to 4 and the Euclidean distance was used as the difference between the patterns. The initial cluster centroid in each cluster was initialized randomly. Examples of the results are shown in Figures 13, 14, and 15. The horizontal axis shows the time when $e_t$ was obtained: note that $T$ was about 250 steps in each environment. The vertical axis shows the cluster number according to which the state pattern was classified. In Figure 13, the state pattern was classified into different classes according to whether it was obtained at a straight wall or corner. The same classification was done in Figure 14, where the patterns obtained at 70-degree corners were classified as class 4. As shown in Figure 15, the patterns obtained at a curved wall were classified as class 2. These results show that our binary pattern could represent the difference in the kind of wall, either a curved wall or a corner angle.

In the above experiments, a designer judged how the robot went around the environments. It is possible for the robot itself to make this judgment itself if a mark is temporarily installed. Now we consider the environmental size and

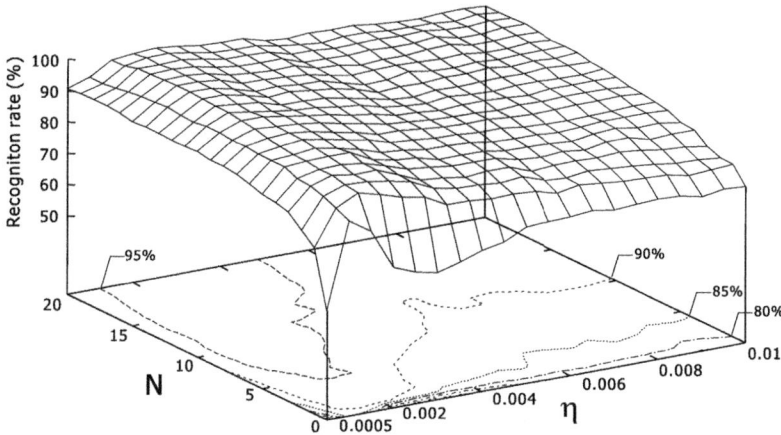

**Fig. 11.** Recognition rates for proposed method using several parameter sets (13 environments)

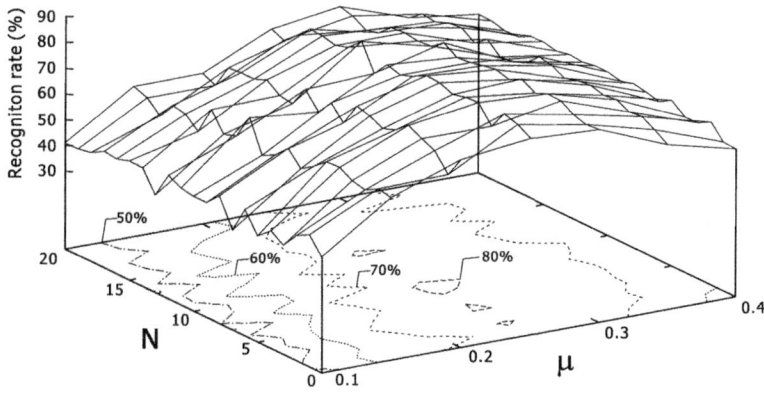

**Fig. 12.** Recognition rates for Wall using several parameter sets (13 environments)

discernment performance. In AEM, the weights are set up for every rule and a model is made using the value accumulated in the order of the rule series. This means that the accumulation-based AEM tends to be influenced by the environment becoming large. On the other hand, our method is not easily affected. Therefore, our method will be more effective than AEM when the environment becomes large.

## 3.4   Discernment with a Random Start Position

We confirmed the effectiveness of our method when the start position was set at random. In this experiment, we set the start position randomly in each trial and obtained the sensory pattern sequence. The matching error was calculated by

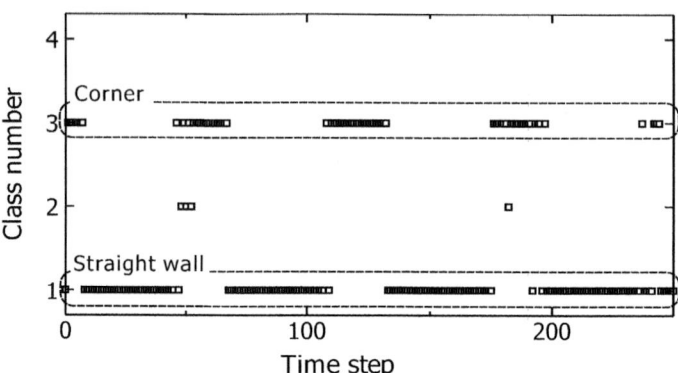

**Fig. 13.** Transition of clusters (env. 7)

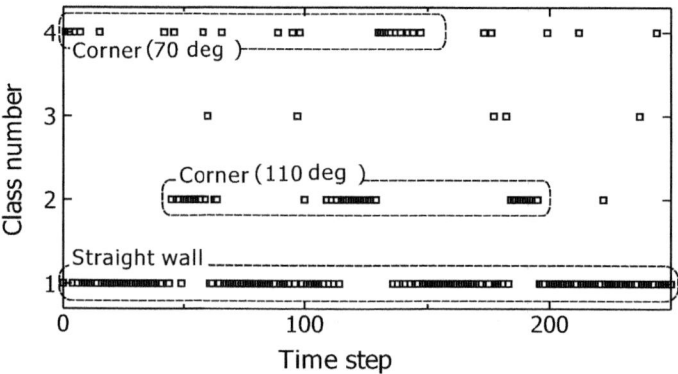

**Fig. 14.** Transition of clusters (env. 10)

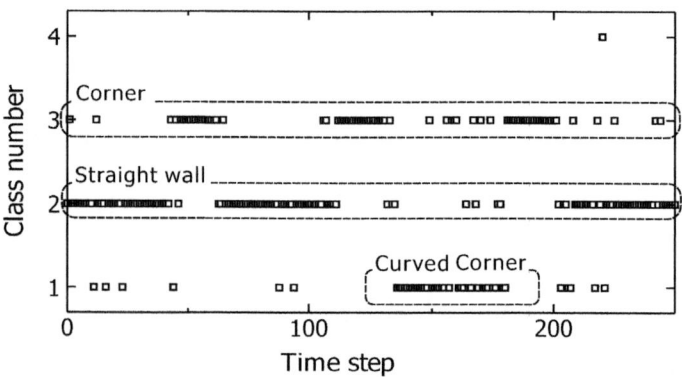

**Fig. 15.** Transition of clusters (env. 13)

**Table 3.** Partial ratio and recognition rates for 13 environments

| Env. | Partial ratio (maximum recog. rate) | |
| :---: | :---: | :---: |
| | Fixed start position | Random start position |
| 1 | 39% (100%) | 100% (76.7%) |
| 2 | 100% (93.3%) | 100% (80.0%) |
| 3 | 10% (100%) | 34% (100%) |
| 4 | 17% (100%) | 20% (100%) |
| 5 | 48% (100%) | 100% (96.7%) |
| 6 | 100% (93.3%) | 100% (93.3%) |
| 7 | 56% (100%) | 56% (100%) |
| 8 | 46% (100%) | 54% (100%) |
| 9 | 72% (100%) | 77% (100%) |
| 10 | 59% (100%) | 60% (100%) |
| 11 | 65% (100%) | 65% (100%) |
| 12 | 51% (100%) | 57% (100%) |
| 13 | 36% (100%) | 37% (100%) |

shifting the sensor patterns by one step at a time and creating a test model; the identification result was then acquired from the minimum error. The experiment was performed for both 7 and 13 environments. Threshold $\eta = 0.003$ and $N = 10$ were used in the 7-environment identification and $\eta = 0.002$ and $N = 14$ were used in the 13-environment identification. The recognition rates for 7- and 13-environment identification were 97.6% and 95.1%, respectively. These results show that our method is also effective when the start position is random.

## 3.5   Discernment by Partial Sequence

When our method is applied to a real environment, it is desirable for it to be able to discriminate from a partial sequence. Therefore, we examined the effectiveness of the method using partial sequence sensory pattern $s_0, \cdots, s_t, \cdots, s_{T_{part}}$. When a partial series is short, since much $e_t$ is set to a zero vector, discernment becomes difficult. Therefore, we calculated the matching error using $T_{part}$ instead of $T_e$ in Equation (4). We carried out the experiment using the 13 environments. The identification was performed for both fixed and random start positions, with $\eta = 0.002$ and $N = 14$ in both cases. In this experiment, we investigated the maximum recognition rate as the length of the partial series was increased. The maximum recognition rate and partial ratio are listed in Table 3. The partial ratio is defined as

$$\text{Partial ratio (\%)} = \frac{\text{Length of partial sequence}}{\text{Length of complete sequence}} \times 100 \qquad (6)$$

With a fixed start position, the average partial ratio and recognition rate were 54% and 99%, respectively. With a random start position, the averages were

66.6% and 95.9%, respectively. When the start position was fixed, we expected our method to achieve high discernment from the data based on about 60% of the length of the sequence for the whole environment in each of these 13 environments. In the random start position case, we expected to get accurate identification from the data based on about 70% of the whole environment sequence length. However, the discernment performance was strongly influenced by the environmental form.

## 4   Conclusions and Future Work

In this paper, we presented a modeling method based on state representation, which represents a change in sensory information. This model enables a mobile robot to identify its environment. Experiments on a real mobile robot having only low-sensitivity infrared sensors showed the effectiveness of our method. A comparison between our method and a conventional one showed that ours had higher performance.

In our method, the size of the robot and the arrangement of the sensors and the sensing area affect the model. We will investigate these effects in the future. Our method assumes that the environment is a closed area. Therefore, it cannot identify an environment where, for example, the landmark is in the center of the room. For such a case, a motion that is not along a wall is needed. Moreover, it cannot handle an object that moves into the environment: for this case, we must propose a state representation based on information about the robot's motion. We plan to investigate these issues in the future.

## Acknowledgment

This research has been supported by funding from the Ministry of Education, Science, Sports and Culture, Grant-in-Aid for Young Scientists (B), 21700219, 2010 and the Kayamori Foundation of Information Science Advancement.

## References

1. Matarić, M.J.: Integration of representation into goal-driven behavior-based robots. IEEE Transactions on Robotics and Automation 8(3), 304–312 (1992)
2. Matsumoto, Y., Inaba, M., Inoue, H.: View-based navigation using an omniview sequence in a corridor environment. Machine Vision and Applications 14, 121–128 (2003)
3. Iwasa, H., Aihara, N., Yokoya, N., Takemura, H.: Memory-based self-localization using omnidirectional images. Systems and Computers in Japan 34(5), 56–68 (2003)
4. Nishida, K., Tanaka, T., Kurita, T.: Constructing a map of place cells for mobile robot navigation. International Congress Series, vol. 1269, pp. 189–192 (2004)
5. Nehmzow, U., Smithers, T.: Mapbuilding using self-organizing networks in really useful robots. In: Proceedings of the First International Conference on Simulation of Adaptive Behavior, pp. 152–159 (1991)

6. Yamada, S., Murota, M.: Unsupervised learning to recognize environments from behavior sequences in a mobile robot. In: Proceedings of IEEE International Conference on Robotics and Automation (ICRA98), pp. 1871–1876 (1998)
7. Duchon, A.P., Kaelbling, L.P., Warren, W.H.: Ecological robotics. Adaptive Behavior 6(3-4), 473–507 (1998)
8. Takahashi, Y., Asada, M., Hosoda, K.: Reasonable performance in less learning time by real roobot based on incremental state space segmentation. In: Proceedings of IEEE/RSJ International Conference on Intelligent Robots and Systems(IROS'96), pp. 1518–1524 (1996)
9. Nakamura, T., Takamura, S., Asada, M.: Behavior-based map representation for a sonor-based mobile robot by statistical methods. In: Proceedings of IEEE/RSJ International Conference on Intelligent Robots and Systems (IROS'96), pp. 276–283 (1996)
10. http://www.e-puck.org/

# Polymorphic Particle Swarm Optimization

Christian Veenhuis

Berlin University of Technology, Berlin, Germany
veenhuis@googlemail.com

**Abstract.** In recent years a swarm-based optimization methodology called Particle Swarm Optimization (PSO) has developed. If one wants to apply PSO one has to specify several parameters as well as to select a neighborhood topology. Several topologies being widely used can be found in literature. This raises the question, which one fits best to your application at hand. To get rid of this topology selection problem, a new concept called Polymorphic Particle Swarm Optimization (PolyPSO) is proposed. PolyPSO generalizes the standard update rule by a polymorphic update rule. The mathematical expression of this polymorphic update rule can be changed on symbolic level. This polymorphic update rule is an adaptive update rule changing symbols based on accumulative histograms and roulette-wheel sampling. PolyPSO is applied to four typical benchmark functions known from literature. In most cases it outperforms the other PSO variants under consideration. Since PolyPSO performs either as best or second best it can be used as alternative to solve this way the topology selection problem. Additionally, PolyPSO significantly outperforms the standard PSO methods in higher dimensional problems.

## 1  Introduction

In recent years a swarm-based optimization methodology called Particle Swarm Optimization (PSO) has developed. If one wants to apply PSO one has to specify several parameters as well as to select a neighborhood topology. Specifying the parameters is not that difficult since there are standard parameters widely used in literature. Additionally, some authors [10,12] introduced adaptive PSO variants, which determine the parameters by themselfs so they don't need to be pre-defined. But only few work is done concerning the topology selection process. In literature, 3 typical topologies can be found being widely used. But, which one of them fits best to your application at hand? All your experiments need to be conducted with each of them to be sure.

In [7,8] the authors solved this problem by incorporating the *gbest* and *lbest* PSO into one algorithm called unified PSO, which is based on Clerc's constriction PSO. They compute a velocity for the *gbest* topology ($G_i^{(t+1)}$) as well as for the *lbest* topology ($L_i^{(t+1)}$). Then, both velocities are unified by a mixed state strategy $u G_i^{(t+1)} + (1 - u) L_i^{(t+1)}$ leading to the next velocity of the particle. Each particle has its own $u$ factor encoded at the end of its position vector. This way, they realized self-adaptivity.

Polymorphism is well-known in nature and refers to the situation that organisms of a species can have more than one phenotype. Computer viruses using polymorphic code to fight against the pattern analysation performed by anti-virus software are another

M.L. Gavrilova et al. (Eds.): Trans. on Comput. Sci. VIII, LNCS 6260, pp. 20–40, 2010.
© Springer-Verlag Berlin Heidelberg 2010

example. These viruses change their code while maintaining the same functionality. In this paper polymorphism is transferred to algorithms, especially the Particle Swarm Optimization method. If the class of PSO algorithms is considered as a species, then different neighborhood topologies and update rules lead to different PSO instances, that is phenotypes, of this algorithm.

The main aim of the proposed concept called Polymorphic Particle Swarm Optimization (PolyPSO) is to get rid of the topology selection process. For this, an adaptive update rule with polymorphic capabilities is defined. This polymorphic update rule replaces the standard update rule in PSO and is capable of mapping all 3 standard PSO variants used in literature. Adaptivity is realized on symbolic level changing the mathematical expression of the update rule based on success rates.

This paper is organized as follows. Section 2 introduces the Particle Swarm Optimization algorithm. The unified PSO is presented in section 3. A concept of polymorphic symbols and equations is introduced in section 4. Based on this, in section 5 the Polymorphic Particle Swarm Optimization approach is defined. In section 6 the conducted experiments with some results are presented. Finally, in section 7 some conclusions are drawn.

## 2 Particle Swarm Optimization

Particle Swarm Optimization (PSO), as introduced by Kennedy and Eberhart [3] [5], is an optimization algorithm modeling the flocking of birds flying around a peak in a landscape. In PSO the birds are substituted by particles and the peak in the landscape is the peak of a fitness function. The particles are flying through the search space forming flocks around peaks of fitness functions.

Let $N_{dim}$ be the dimension of the problem (i.e., the dimension of the search space $\mathbb{R}^{N_{dim}}$), $N_{part}$ the number of particles and $\mathcal{P}$ the set of particles $\mathcal{P} = \{P_1, ..., P_{N_{part}}\}$. Each particle $P_i = (x_i, v_i, p_i)$ has a current position in the search space ($x_i \in \mathbb{R}^{N_{dim}}$), a velocity ($v_i \in \mathbb{R}^{N_{dim}}$) and the personally best found position in history ($p_i \in \mathbb{R}^{N_{dim}}$), i.e., the own experience of this particle.

In PSO, the set of particles $\mathcal{P}$ is initialized at time step $t = 0$ with randomly created particles $P_i^{(0)}$. The initial $p_i$ are set to the corresponding initial $x_i$. Then, for each time step $t$, the next position $x_i^{(t+1)}$ and velocity $v_i^{(t+1)}$ of each particle $P_i^{(t)}$ are computed as shown in Eqns. (1) and (2).

$$v_i^{(t+1)} = w_I^{(t)} v_i^{(t)} + (w_P r_P) \odot (p_i^{(t)} - x_i^{(t)}) + (w_N r_N) \odot (n_i^{(t)} - x_i^{(t)}) \qquad (1)$$

$$x_i^{(t+1)} = x_i^{(t)} + v_i^{(t+1)} \qquad (2)$$

Here, $r_P$ and $r_N$ are random vectors in $[0, 1]^{N_{dim}}$ (vector to vector multiplication $\odot$ is realized component-wise). The personally best found position of the best neighbor particle at time $t$ is denoted with $n_i^{(t)} \in \mathbb{R}^{N_{dim}}$.

As presented in [5] [6] [2] there are several possibilities to define the neighborhood of a particle. The widely used ones are depicted in Figure 1. The global (*gbest*) PSO uses a star topology [4]. This way, the whole swarm is the neighborhood of each particle. For the local (*lbest*) PSO a ring topology is used [4]. The neighbors of a particle are only the ones at both sides up to a specified radius. A radius of $r$ would mean $r$ neighbors of the left and $r$ neighbors of the right side. The global and local PSO use a pre-defined static neighborhood that does not change during the run. In [11] a dynamic neighborhood is used. There, a distance function between particles determines the neighborhood. In this paper the $k$ nearest particles in search space according to Euclidian distance are used as neighbors. A PSO using this topology shall be called distance PSO.

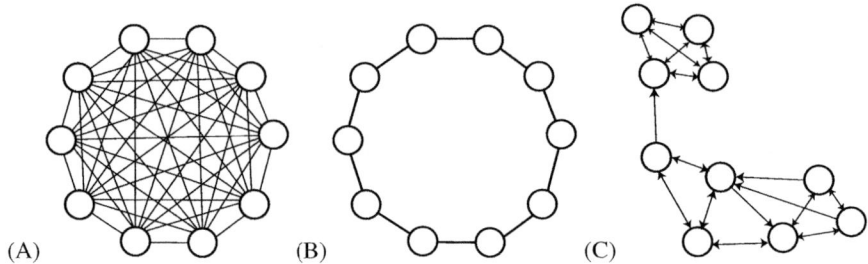

**Fig. 1.** Widely used neighborhood topologies include star (A), ring (B) and distance based (C). The distance based example uses the $k = 3$ nearest neighbors.

The inertia weight $w_I^{(t)}$ determines the influence of the particle's own velocity, i.e., it represents the confidence of the particle to its own position (typically $w_I \in [0.1, 1.0]$). To improve the convergence behavior, this weight is decreased over time [9] [5] by an amount of $a_I$ (typically $a_I = 0.001$). $w_P$ is the influence of the personally best position found so far. The influence of the best particle of the neighborhood is denoted by $w_N$.

To avoid chaotic behavior, the new velocity $v_i^{(t+1)}$ is clamped to a pre-defined interval $[-V_{max}, +V_{max}]$. Also, the new position $x_i^{(t+1)}$ is clamped to the problem-specific range $[X_{min}, X_{max}]$. Often, $X_{min}$ is set to $-X_{max}$.

The fitness of a particle is determined by a fitness function $F : \mathbb{R}^{N_{dim}} \to \mathbb{R}$. If the new position $x_i^{(t+1)}$ has a better fitness than the best solution found so far by particle $P_i$, it is stored in memory as shown in Eq. (3) (in case of minimization).

$$p_i^{(t+1)} = \begin{cases} x_i^{(t+1)} & , F(x_i^{(t+1)}) < F(p_i^{(t)}) \\ p_i^{(t)} & , otherwise \end{cases} \qquad (3)$$

The best solution of the run is found at particle $P_b$ with the best personal solution $p_b$. Best solution $p_b$ is always element of the set of all best solutions $\{p_i\}, \forall i \in \{1, \cdots, N_{part}\}$. The best fitness value is $F(p_b) = \min_{i \in \{1, \cdots, N_{part}\}} \{F(p_i)\}$.

## 3   Unified Particle Swarm Optimization

To get rid of the necessity to select a topology, Parsopoulos and Vrahatis proposed in [7,8] a PSO, which incorporates the *gbest* and *lbest* PSO into one algorithm they called unified PSO. The unified PSO is based on the constriction PSO as defined by Clerc and Kennedy in [1]. The main idea is to compute a velocity for the *gbest* topology ($G_i^{(t+1)}$) as well as for the *lbest* topology ($L_i^{(t+1)}$):

$$G_i^{(t+1)} = \chi \, [ \, w_I^{(t)} v_i^{(t)} + (c_1 r_1) \odot (p_i^{(t)} - x_i^{(t)}) + (c_2 r_2) \odot (n_{i,global}^{(t)} - x_i^{(t)}) \, ] \qquad (4)$$

$$L_i^{(t+1)} = \chi \, [ \, w_I^{(t)} v_i^{(t)} + (c_1 r_1) \odot (p_i^{(t)} - x_i^{(t)}) + (c_2 r_2) \odot (n_{i,local}^{(t)} - x_i^{(t)}) \, ] \qquad (5)$$

The best neighbors out of the appropriate neighborhoods are denoted by $n_{i,global}^{(t)}$ and $n_{i,local}^{(t)}$. Furthermore, $\chi$ represents the constriction factor. As in standard PSO (section 2), $c_1$ and $c_2$ represent the cognitive and social coefficients ($w_P$ and $w_N$ in Eq. (1)). Also, the random vectors $r_1$ and $r_2$ are in $[0,1]^{N_{dim}}$ ($r_P$ and $r_N$ in Eq. (1)).

After computing both velocities, they are unified by a mixed state strategy leading to the next velocity of the particle:

$$v_i^{(t+1)} = u G_i^{(t+1)} + (1 - u) L_i^{(t+1)} \qquad (6)$$

The unification factor $u$ is in $[0,1]$ and balances the influence of the *gbest* and *lbest* topology. Note that both standard PSO variants are special cases. If $u = 1$ one gets the global PSO. The local PSO is obtained with $u = 0$.

Each particle has its own unification factor encoded at the end of its position vector realizing this way self-adaptivity:

$$x_i = ( \, x_{i_1} \, , \, x_{i_2} \, , \, \cdots \, , \, x_{i_{N_{dim}}} \, , \, u \, )^T \qquad (7)$$

## 4  Polymorphic Symbols

In this work a symbol means each object or variable used in terms or equations as for instance $x$, $y$ and $z$ in $x + y = z$. A *Polymorphic Symbol* is a more abstract symbol, a 'meta-symbol', representing a finite set of symbols. To avoid term confusions, symbols being used in terms or equations are called variables, hereafter. Let $V = < v_1, \cdots, v_N >$ denote an ordered list of $N$ variables $v_i$. Furthermore, let $P = < p_1, \cdots, p_N >$ denote an ordered list of properties, whereby property $p_i$ corresponds to variable $v_i$.

**Definition 1 (Polymorphic Symbol).** *A polymorphic symbol $S = (V_S, P_S, C_S)$ is comprised of a list of variables $V_S$, their corresponding properties $P_S$ as well as a determination procedure $C_S : \{S_i\} \mapsto \{v_1, \cdots, v_N\}$. Procedure $C_S$ gets a polymorphic symbol and returns a valid variable out of $V_S$. A variable $C_S(S)$ is called a configuration of $S$.*

**Example.** Lets define a polymorphic symbol

$$\mathcal{P} \; = \; ( \, V_{\mathcal{P}} = < x, y, z > , \, P_{\mathcal{P}} = < 1, 0, 0 > , \, C_{\mathcal{P}} \, ).$$

Furthermore, lets determine a variable by

$$C_{\mathcal{P}}(\mathcal{P}) = \arg \max_{v_i \in \{v_1, \cdots, v_N\}} (p_i).$$

This procedure returns the variable whose corresponding property has the highest value. The configuration of $\mathcal{P}$ can now be controlled by manipulating its properties $P_{\mathcal{P}}$. If we set $P_{\mathcal{P}} = < 1,0,0 >$, then $C_{\mathcal{P}}(\mathcal{P}) = v_1 = x$. For $P_{\mathcal{P}} = < 0,1,0 >$ we obtain $C_{\mathcal{P}}(\mathcal{P}) = v_2 = y$. Finally, $P_{\mathcal{P}} = < 0,0,1 >$ leads to $C_{\mathcal{P}}(\mathcal{P}) = v_3 = z$.

**Polymorphic Equations.** It is possible to define *Polymorphic Equations*, if at least one variable is replaced by a polymorphic symbol. Lets take the example equation

$$x + y = z$$

and lets replace the first variable by the configuration of polymorphic symbol $\mathcal{P}$:

$$C_{\mathcal{P}}(\mathcal{P}) + y = z$$

Depending on $P_{\mathcal{P}}$ three different equations are now possible:

$$x + y = z$$
$$y + y = z$$
$$z + y = z$$

If all variables are replaced by polymorphic symbols $\mathcal{P}_1$, $\mathcal{P}_2$ and $\mathcal{P}_3$ (defined similarily to $\mathcal{P}$) we get

$$C_{\mathcal{P}}(\mathcal{P}_1) + C_{\mathcal{P}}(\mathcal{P}_2) = C_{\mathcal{P}}(\mathcal{P}_3)$$

with $3^3 = 27$ different equation configurations. Note that the number of equation configurations is not necessarily the number of different mathematical expressions. Lets assume two different configurations

**a)** $C_{\mathcal{P}}(\mathcal{P}_1) = x$, $C_{\mathcal{P}}(\mathcal{P}_2) = y$, $C_{\mathcal{P}}(\mathcal{P}_3) = z$
**b)** $C_{\mathcal{P}}(\mathcal{P}_1) = y$, $C_{\mathcal{P}}(\mathcal{P}_2) = x$, $C_{\mathcal{P}}(\mathcal{P}_3) = z$

leading to the equations $x + y = z$ and $y + x = z$, respectively. Iff the $+$ operator used is commutative, then both equation configurations lead to the same mathematical expression.

The introduced concept of polymorphic symbols / equations allows for adaptive equations. A method is able to adapt its equations by itself by changing the properties of the used polymorphic symbols.

## 5    Polymorphic Particle Swarm Optimization

The Polymorphic Particle Swarm Optimization (PolyPSO) concept replaces the typically used update rule by a polymorphic update rule. In PSO, the movement of a particle is guided by a cognitive term $(w_P r_P) \odot (p_i^{(t)} - x_i^{(t)})$ as well as a social term $(w_N r_N) \odot (n_i^{(t)} - x_i^{(t)})$. These terms are generalized by a polymorphic term

$$\mathcal{T}_k := (w_k r_k) \odot (C(\mathcal{P}_k) - x_i^{(t)})$$

whereby $w_k$ is the acceleration coefficient, $r_k$ a random vector in $[0,1]^{N_{dim}}$ (again, $\odot$ is realized component-wise) and $\mathcal{P}_k$ a polymorphic symbol.

To solve the topology selection problem, all three neighborhood topologies as presented in section 2 need to be provided. The $n_i^{(t)}$ variable of the social term is replaced by three variables representing the best neighbor out of the appropriate neighborhoods: $n_{i,global}^{(t)}$, $n_{i,local}^{(t)}$ and $n_{i,distance}^{(t)}$. The used topology parameters $r = 1$ and $k = 2$ are set to the default values.

Let $\mathcal{P}_1$, $\mathcal{P}_2$, $\mathcal{P}_3$ and $\mathcal{P}_4$ be polymorphic symbols defined as:

$$\mathcal{P}_k := (V, P_{\mathcal{P}_k}, C) \tag{8}$$

$$V = \; < n_{i,global}^{(t)}, n_{i,local}^{(t)}, n_{i,distance}^{(t)}, p_i^{(t)}, x_i^{(t)} >$$

$$P_{\mathcal{P}_k} = \; < 1, 1, 1, 1, 1 >$$

Each $\mathcal{P}_k$ uses the same variables and determination procedure. But each $\mathcal{P}_k$ has its own distinct property $P_{\mathcal{P}_k}$, since all variables are meant to be independently modifiable.

In PolyPSO the properties $P_{\mathcal{P}_k}$ are interpreted as accumulative histograms. To get the configurations, these histograms are sampled using roulette-wheel sampling. That is, $C(\mathcal{P}_k)$ returns variable $v_j$ with probability $\frac{(P_{\mathcal{P}_k})_j}{\Sigma_s (P_{\mathcal{P}_k})_s}$. Thus, the higher a histogram bin $(P_{\mathcal{P}_k})_j$, the higher the probability that the corresponding variable $v_j$ is the configuration determined for a given polymorphic symbol.

Based on the four polymorphic symbols $\mathcal{P}_1, \cdots, \mathcal{P}_4$ we get four polymorphic terms $\mathcal{T}_1 = (w_1 r_1) \odot (C(\mathcal{P}_1) - x_i^{(t)}), \cdots, \mathcal{T}_4 = (w_4 r_4) \odot (C(\mathcal{P}_4) - x_i^{(t)})$. With them, the polymorphic update rule can now be defined as:

$$v_i^{(t+1)} = w_I^{(t)} v_i^{(t)} + \mathcal{T}_1 + \mathcal{T}_2 + \mathcal{T}_3 + \mathcal{T}_4 \tag{9}$$

To allow update rules using all influences (all three topologies together with own experience), four polymorphic terms are used. But the list of variables $V$ also contains the current position $x_i^{(t)}$ of the particle considered. This allows to eliminate the influence of a polymorphic term, if the configuration of its polymorphic symbol returns $x_i^{(t)}$ $(C(\mathcal{P}_k) = x_i^{(t)} \Rightarrow (w_k r_k) \odot (C(\mathcal{P}_k) - x_i^{(t)}) = 0)$. Thus, the actual number of used terms is 0 - 4 and can be controlled by PolyPSO itself.

Note that the three standard PSOs are special cases of this polymorphic update rule. If one sets $P_{\mathcal{P}_1} = < 0, 0, 0, 1, 0 >$ and $P_{\mathcal{P}_3} = P_{\mathcal{P}_4} = < 0, 0, 0, 0, 1 >$, then one gets the global PSO with $P_{\mathcal{P}_2} = < 1, 0, 0, 0, 0 >$, the local PSO with $P_{\mathcal{P}_2} = < 0, 1, 0, 0, 0 >$ and the distance PSO with $P_{\mathcal{P}_2} = < 0, 0, 1, 0, 0 >$.

**Initialization.** PolyPSO is initialized the same way as standard PSO (see section 2). Additionally, all bins of the accumulative histograms of all polymorphic symbols are set to 1 ($P_{\mathcal{P}_1} = P_{\mathcal{P}_2} = P_{\mathcal{P}_3} = P_{\mathcal{P}_4} = < 1, 1, 1, 1, 1 >$). This way all symbols have the same probability to be chosen.

**Iteration.** Basically, the iteration is the same as in standard PSO. What is different is the integration of the accumulative histograms as well as the polymorphic update rule. Algorithm 1 outlines the whole iteration step: The first step is to initialize the histograms for accumulating successful configurations denoted as $S_{\mathcal{P}_1}, \cdots, S_{\mathcal{P}_4}$. Then, all particles of the current generation are moved based on the polymorphic update rule in Eq. (9). After updating the global best particle $g$ and the particle's personal best (experience), the best neighbors of all three neighborhood topologies are determined. These three neighbors $(n_{i,global}^{(t)}, n_{i,local}^{(t)}, n_{i,distance}^{(t)})$ together with the current particle $x_i^{(t)}$ and its experience $p_i^{(t)}$ are needed for the polymorphic symbols (remember the definition of $V$ in Eq. (8)). Afterwards, the actual configuration of all 4 polymorphic symbols are determined using $C(\cdot)$. These are hold in variables $c_1, \cdots, c_4$ for the case of success to be able to count the right symbols (a second call to $C(\cdot)$ could deliver another symbol caused by the sampling process). Then, the new position $x_i^{(t+1)}$ is computed using the polymorphic update rule. If this new position is better than the current, then the successful configuration $c_1, \cdots, c_4$ used for the polymorphic update rule is counted. For this, the appropriate histogram bins of the success histograms $S_{\mathcal{P}_1}, \cdots, S_{\mathcal{P}_4}$ are incremented. To determine the index number of the bin for a given symbol in configuration $c_i$, the *indexof*$(\cdot)$ operator is used. This operator returns 1 if $c_i = n_{i,global}^{(t)}$, 2 if $c_i = n_{i,local}^{(t)}$ and so on. Finally, after all particles were processed, the success histograms are added to the accumulative histograms $P_{\mathcal{P}_1}, \cdots, P_{\mathcal{P}_4}$. This accumulates all successful configurations over all iterations realizing this way the adaptivity. The last step is to update the inertia weight $w_I$.

## 6   Experiments

To evaluate the capabilities of PolyPSO, four typical benchmark functions as known from literature (Sphere, Rosenbrock, Rastrigin, Griewank), representing all four combinations of unimodal/multimodal with/without dependencies between the variables, were chosen.

For each of these benchmark functions, the PolyPSO approach, the UnifiedPSO as well as the standard PSO method with all 3 neighborhood topologies were conducted. We need to compare to all topologies, since our main aim is to get rid of the topology selection process. This makes only sense, if PolyPSO has at least a comparable performance compared to all topologies considered.

To get an impression of the performance, experiments were conducted for the lower dimensions $N_{dim} = 10, 20, \cdots, 100$ as well as for the higher dimensions $N_{dim} = 100, 200, \cdots, 500$. For each benchmark function and dimension, 100 independent runs with 1000 iterations were performed. All methods used the standard parameters: the number of particles $N_{part}$ is set to 20, the range $[X_{min}, X_{max}]$ is set according to the benchmark functions (see sub-sections), the inertia weight is decreased from 1.0 to 0.1 over all 1000 iterations ($a_I = \frac{1-0.1}{1000}$), the radius of the ring topology $r$ is set to 1 and the $k = 2$ nearest neighbors are used for the distance based topology. Furthermore, the weights of the polymorphic terms $w_1, \cdots, w_4$ were all set to 0.5. As typical in standard PSO, $V_{max}$ is set as sum of all weights, i.e., $V_{max} = 2$. For the UnifiedPSO approach the above standard parameters are used, too. The constriction parameters are set as proposed by the authors in [7]: the constriction factor $\chi = 0.729$ and both coefficients are set to $c_1 = c_2 = 2.05$.

---

**Algorithm 1.** Iteration of PolyPSO

---

1: // *Initialize histograms to count successful polymorphic symbols*
2: $S_{\mathcal{P}_1} \leftarrow <0,0,0,0,0>$
3: $S_{\mathcal{P}_2} \leftarrow <0,0,0,0,0>$
4: $S_{\mathcal{P}_3} \leftarrow <0,0,0,0,0>$
5: $S_{\mathcal{P}_4} \leftarrow <0,0,0,0,0>$
6:
7: // *Update each particle of swarm*
8: **for all** $P_i^{(t)} \in \mathcal{P}$ **do**
9:
10:     // *Update global best g*
11:     **if** $F(x_i^{(t)}) < F(g)$ **then**
12:         $g \leftarrow x_i^{(t)}$
13:     **end if**
14:
15:     // *Update personal best* $p_i^{(t)}$
16:     **if** $F(x_i^{(t)}) < F(p_i^{(t)})$ **then**
17:         $p_i^{(t)} \leftarrow x_i^{(t)}$
18:     **end if**
19:
20:     Determine neighbors $n_{i,global}^{(t)}, n_{i,local}^{(t)}, n_{i,distance}^{(t)}$
21:
22:     $c_1 = C(\mathcal{P}_1)$                    // *Determine configurations of polymorphic symbols*
23:     $c_2 = C(\mathcal{P}_2)$
24:     $c_3 = C(\mathcal{P}_3)$
25:     $c_4 = C(\mathcal{P}_4)$
26:
27:     $r_1, r_2, r_3, r_4 \sim U(0,1)^{N_{dim}}$        // *Determine random vectors*
28:
29:     $v_i^{(t+1)} \leftarrow w_I^{(t)} v_i^{(t)} + (w_1 r_1) \odot (c_1 - x_i^{(t)}) + (w_2 r_2) \odot (c_2 - x_i^{(t)}) + (w_3 r_3) \odot (c_3 - x_i^{(t)}) + (w_4 r_4) \odot (c_4 - x_i^{(t)})$
30:
31:     $x_i^{(t+1)} \leftarrow x_i^{(t)} + v_i^{(t+1)}$
32:
33:     // *Count configuration, if successful*
34:     **if** $F(x_i^{(t+1)}) < F(x_i^{(t)})$ **then**
35:         $(S_{\mathcal{P}_1})_{indexof(c_1)} \leftarrow (S_{\mathcal{P}_1})_{indexof(c_1)} + 1$        // *Update success histograms*
36:         $(S_{\mathcal{P}_2})_{indexof(c_2)} \leftarrow (S_{\mathcal{P}_2})_{indexof(c_2)} + 1$
37:         $(S_{\mathcal{P}_3})_{indexof(c_3)} \leftarrow (S_{\mathcal{P}_3})_{indexof(c_3)} + 1$
38:         $(S_{\mathcal{P}_4})_{indexof(c_4)} \leftarrow (S_{\mathcal{P}_4})_{indexof(c_4)} + 1$
39:     **end if**
40:
41: **end for**
42:
43: // *Accumulate success histograms*
44: $P_{\mathcal{P}_1} \leftarrow P_{\mathcal{P}_1} + S_{\mathcal{P}_1}$
45: $P_{\mathcal{P}_2} \leftarrow P_{\mathcal{P}_2} + S_{\mathcal{P}_2}$
46: $P_{\mathcal{P}_3} \leftarrow P_{\mathcal{P}_3} + S_{\mathcal{P}_3}$
47: $P_{\mathcal{P}_4} \leftarrow P_{\mathcal{P}_4} + S_{\mathcal{P}_4}$
48:
49: **if** $w_I^{(t)} > 0.1$ **then**
50:     $w_I^{(t+1)} \leftarrow w_I^{(t)} - a_I$
51: **end if**

---

## 6.1  Sphere

The **Sphere** benchmark (see Figure 2) is the following simple unimodal function without dependencies between the variables:

$$f(<x_1,\cdots,x_n>) = \sum_{i=1}^{n} x_i^2$$

$$(X_{min},X_{max}) := (-5.12, 5.12)$$

Global minimum  :  $f(<0,\cdots,0>) = 0$

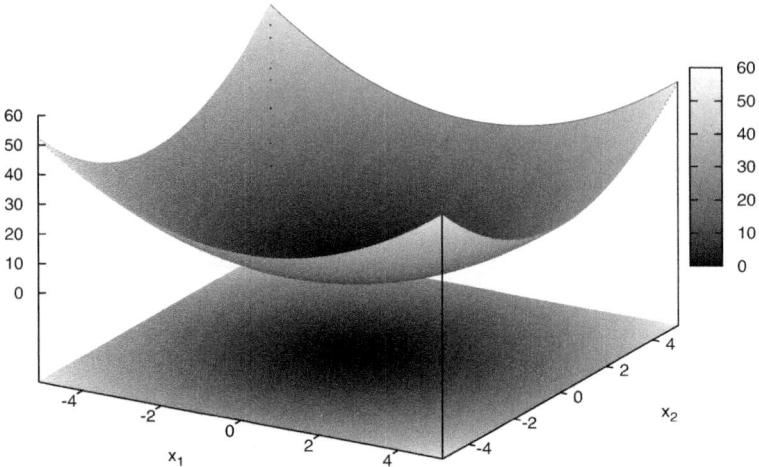

**Fig. 2.** The **Sphere** benchmark for two dimensions

Figure 6 (page 31) reveals that Sphere is best solved by the PolyPSO approach closely followed by Local-PSO. The third best is the UnifiedPSO, whereby all three methods perform comparable up to approx. 35 dimensions. After 70 dimensions, Poly PSO starts to continuously get better than all other methods. As can be seen in Fig. 7 (page 32), approx. at dimension 110, the UnifiedPSO gets better than the Local-PSO in higher dimensions and starts to outperform all standard PSOs.

According to the standard deviation, the Local-PSO performs best followed by Poly PSO for lower dimensions. UnifiedPSO lies in the middle field. For higher dimensions, the PolyPSO approach is the best performer followed by Local-PSO. Interestingly, the higher the dimension, the worse UnifiedPSO's standard deviation.

## 6.2  Rosenbrock

**Rosenbrock**'s benchmark, as depicted in Figure 3, is also a unimodal function, but it is a bit more difficult as Sphere, because there are some dependencies between the variables. It is defined as follows:

$$f(<x_1,\cdots,x_n>) = \sum_{i=1}^{n-1} (100 \cdot (x_{i+1} - x_i^2)^2 + (x_i - 1)^2)$$

$$(X_{min},X_{max}) := (-2.048, 2.048)$$

Global minimum  :  $f(<1,\cdots,1>) = 0$

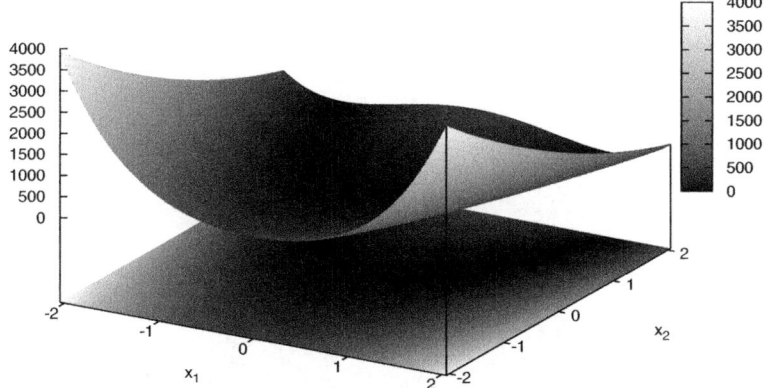

**Fig. 3.** The **Rosenbrock** benchmark for two dimensions

As presented in Figs. 8 (page 33) and 9 (page 34) Rosenbrock is best solved by UnifiedPSO followed by PolyPSO for the lower as well as for the higher dimensions. For the first 58 dimensions, Local-PSO is the second best in lower dimensions. After this it becomes the third best and is outperformed by Global-PSO in higher dimensions.

In the lower dimensions, the UnifiedPSO has the best standard deviation followed by Local-PSO and PolyPSO. But the higher the dimension, the better Global-PSO gets. Starting at approx. 310 dimensions the best performers are Global-PSO, UnifiedPSO and PolyPSO (in this order). Interestingly, the ranking of PolyPSO is mostly in the middle field over all dimensions. Also, UnifiedPSO outperforms PolyPSO over all dimensions.

### 6.3  Rastrigin

**Rastrigin**'s benchmark (shown in Figure 4) is the following multimodal function without dependencies between the variables:

$$f(< x_1, \cdots, x_n >) = 10 \cdot n + \sum_{i=1}^{n} (x_i^2 - 10 \cdot \cos(2 \cdot \pi \cdot x_i))$$

$$(X_{min}, X_{max}) := (-5.12, 5.12)$$

$$\text{Global minimum} \ : \ f(< 0, \cdots, 0 >) = 0$$

As depicted in Figs. 10 (page 35) and 11 (page 36), PolyPSO is the winner and UnifiedPSO the second best for lower as well as for higher dimensions. Among the standard PSOs, the Global-PSO performs best over all dimensions. With increasing dimensions, the Distance-PSO gets slightly better than the Local-PSO.

The best standard deviation in lower dimensions have the PolyPSO and UnifiedPSO methods (in this order). In higher dimensions, mostly the PolyPSO approach performs best. At approx. 380 dimensions the Distance-PSO and in particular the Local-PSO get slightly better than PolyPSO. As with the Sphere benchmark, the higher the dimension, the worse UnifiedPSO's standard deviation.

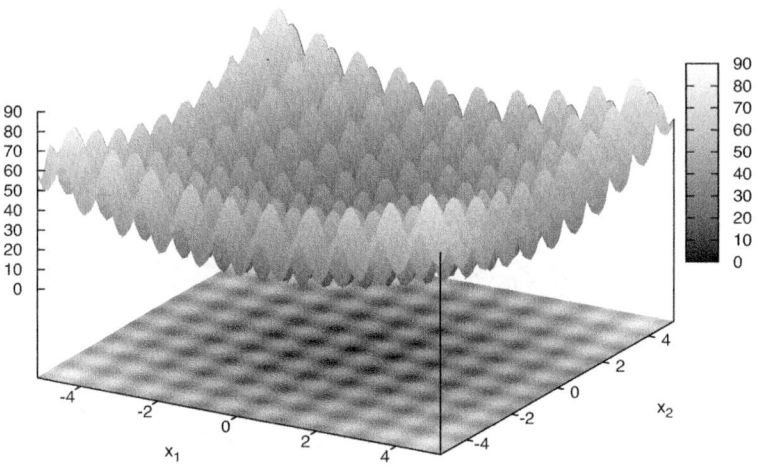

**Fig. 4.** The **Rastrigin** benchmark for two dimensions

## 6.4   Griewank

**Griewank**'s benchmark (see Figure 5) is a multimodal function with strong dependencies between the variables and is defined as follows:

$$f(<x_1, \cdots, x_n>) = 1 + \sum_{i=1}^{n} \frac{x_i^2}{4000} - \prod_{i=1}^{n} \cos(\frac{x_i}{\sqrt{i}})$$

$$(X_{min}, X_{max}) := (-600, 600)$$

Global minimum :   $f(<0, \cdots, 0>) = 0$

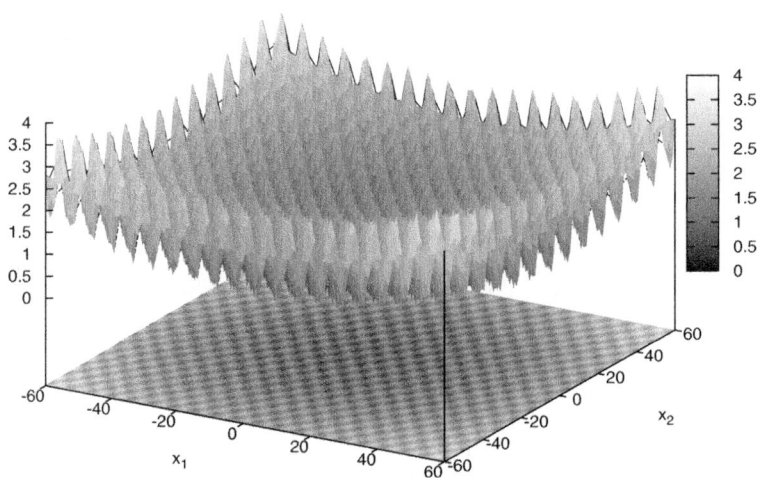

**Fig. 5.** The **Griewank** benchmark for two dimensions

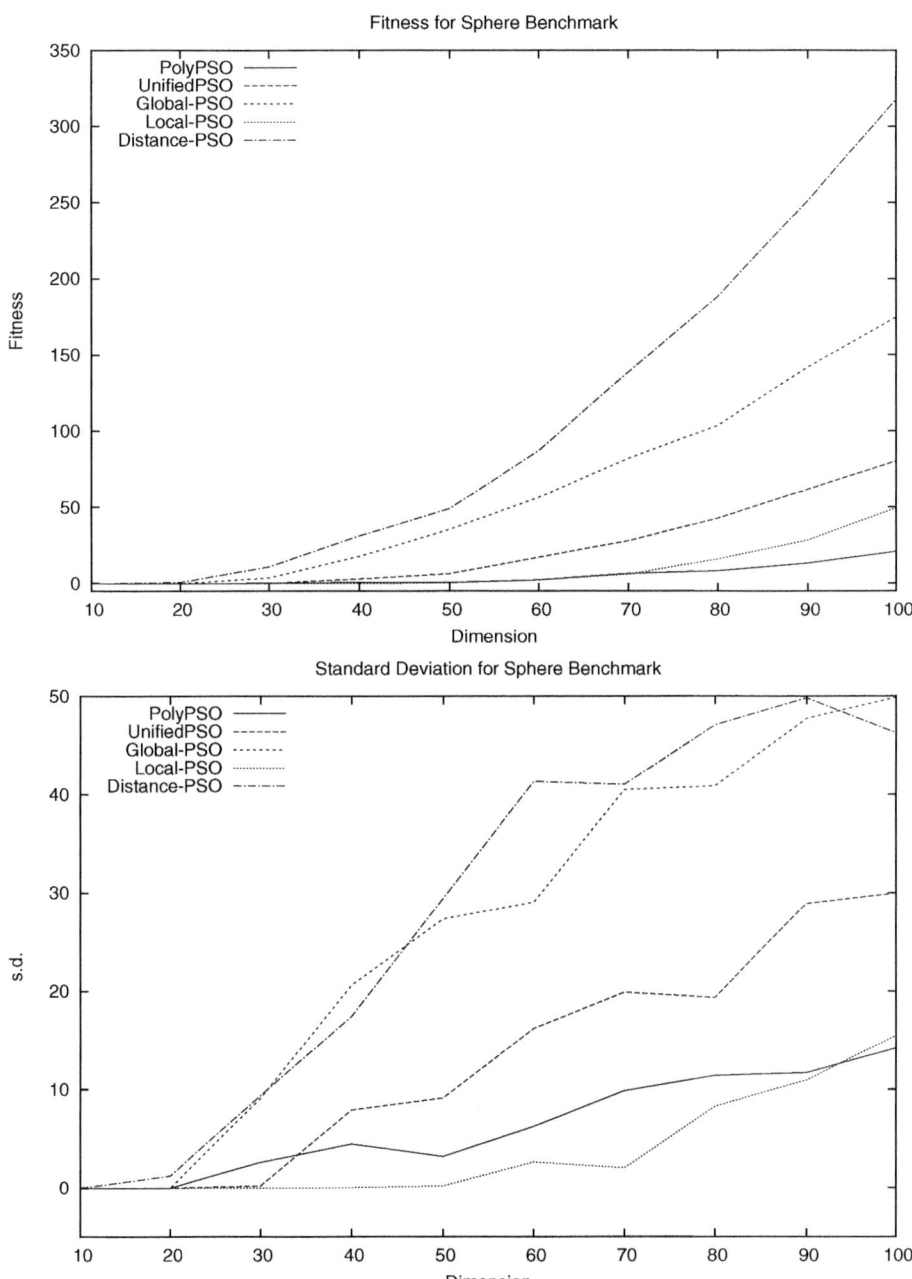

**Fig. 6.** The fitness and standard deviation for the **Sphere** Benchmark over all **low** dimensions

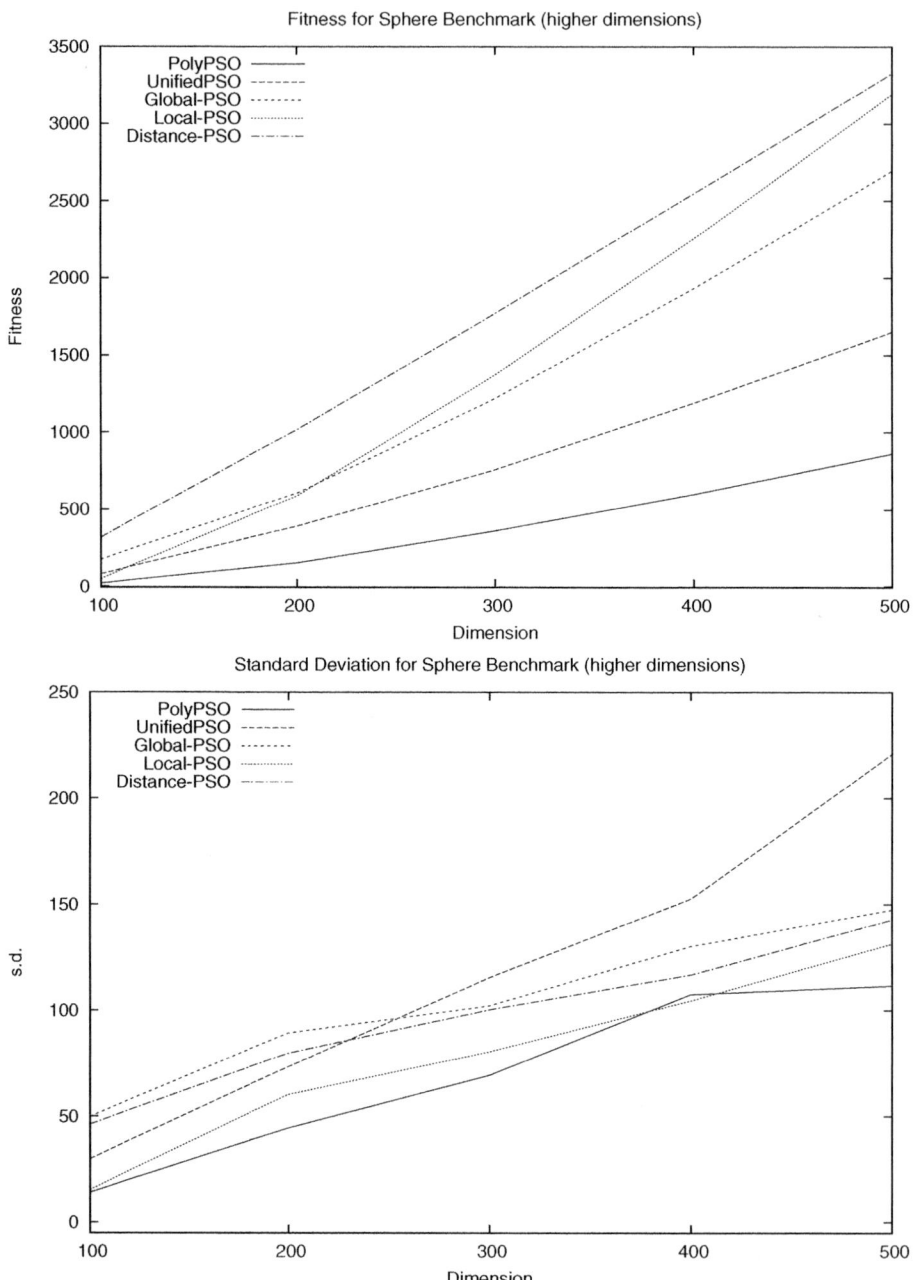

**Fig. 7.** The fitness and standard deviation for the **Sphere** Benchmark over all **high** dimensions

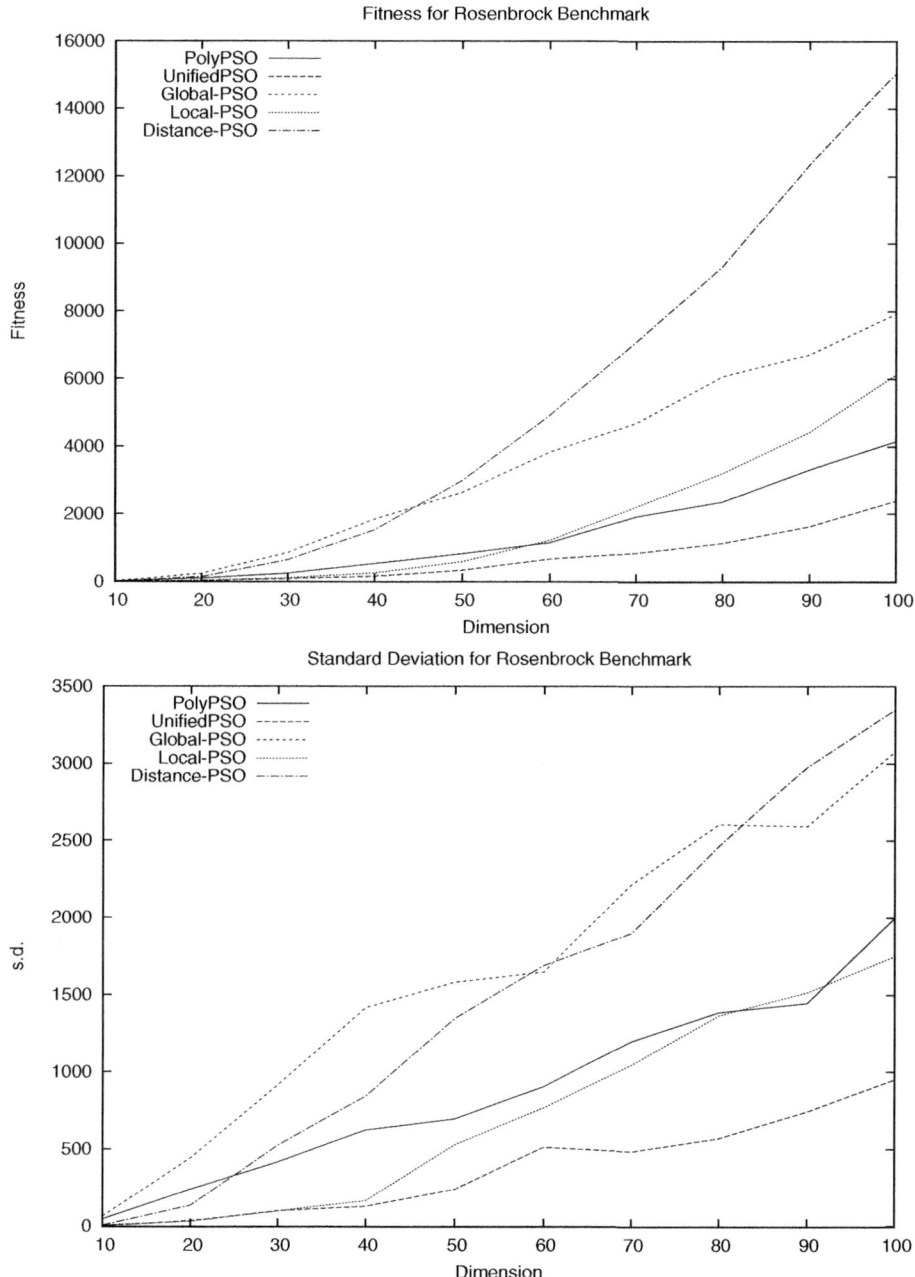

**Fig. 8.** The fitness and standard deviation for the **Rosenbrock** Benchmark over all **low** dimensions

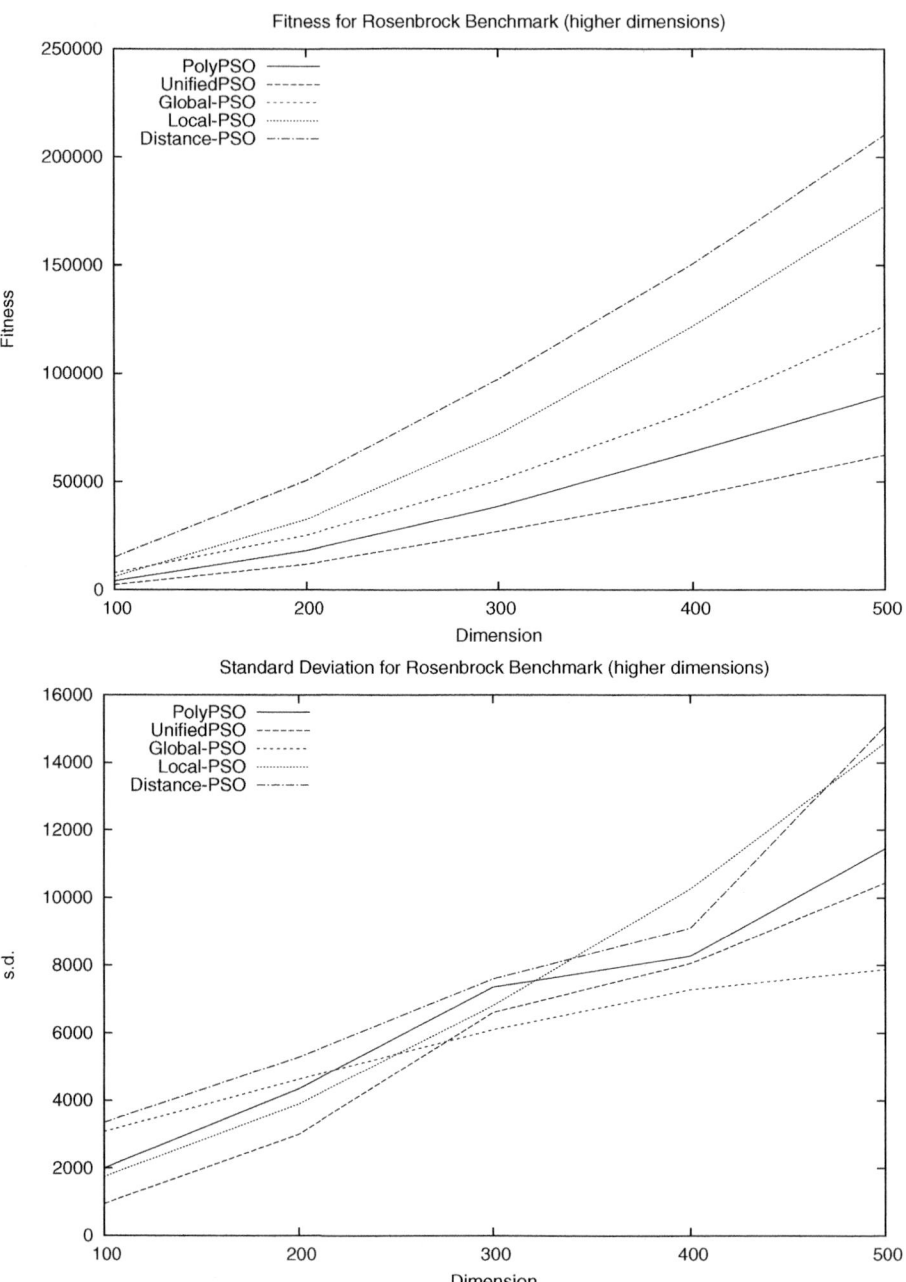

**Fig. 9.** The fitness and standard deviation for the **Rosenbrock** Benchmark over all **high** dimensions

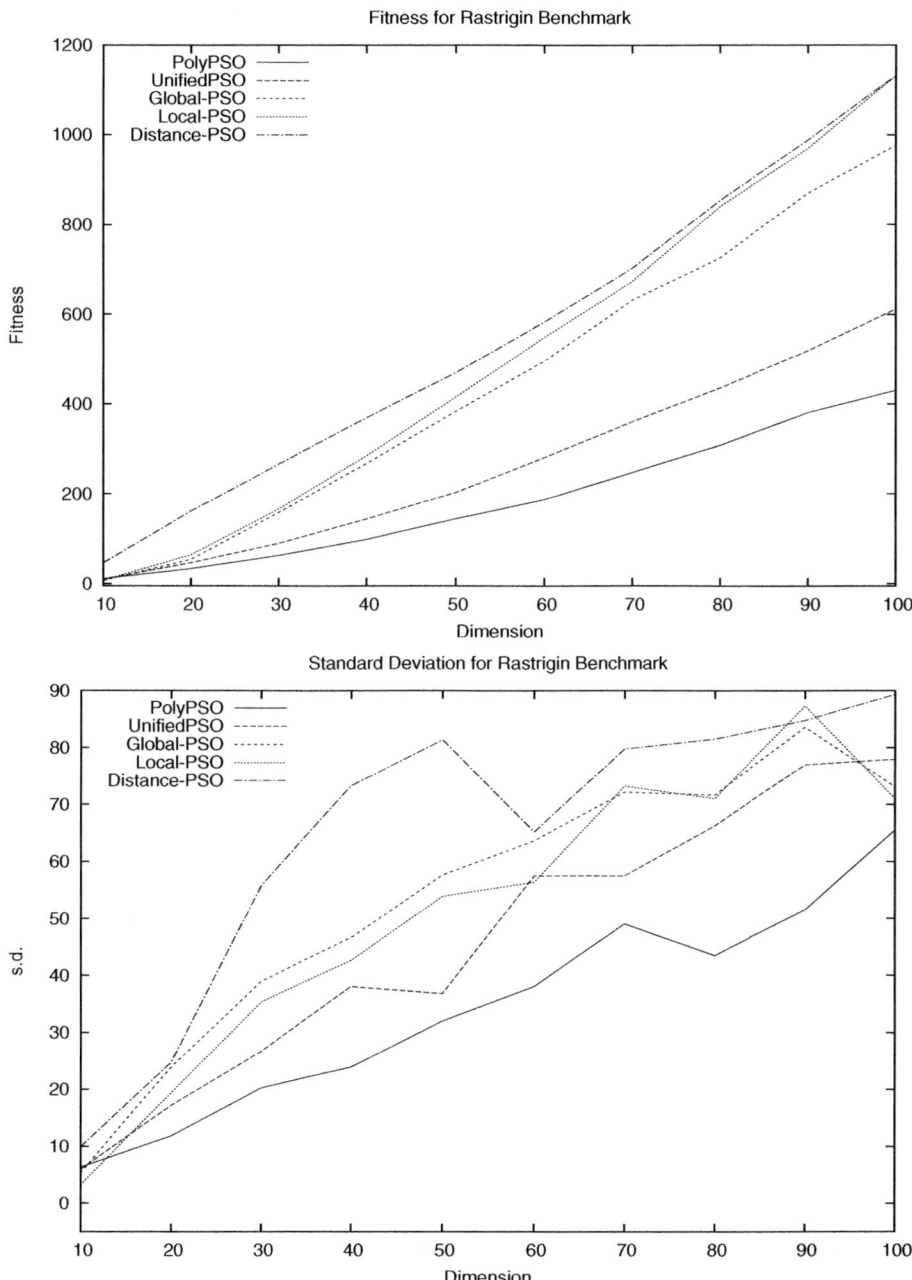

**Fig. 10.** The fitness and standard deviation for the **Rastrigin** Benchmark over all **low** dimensions

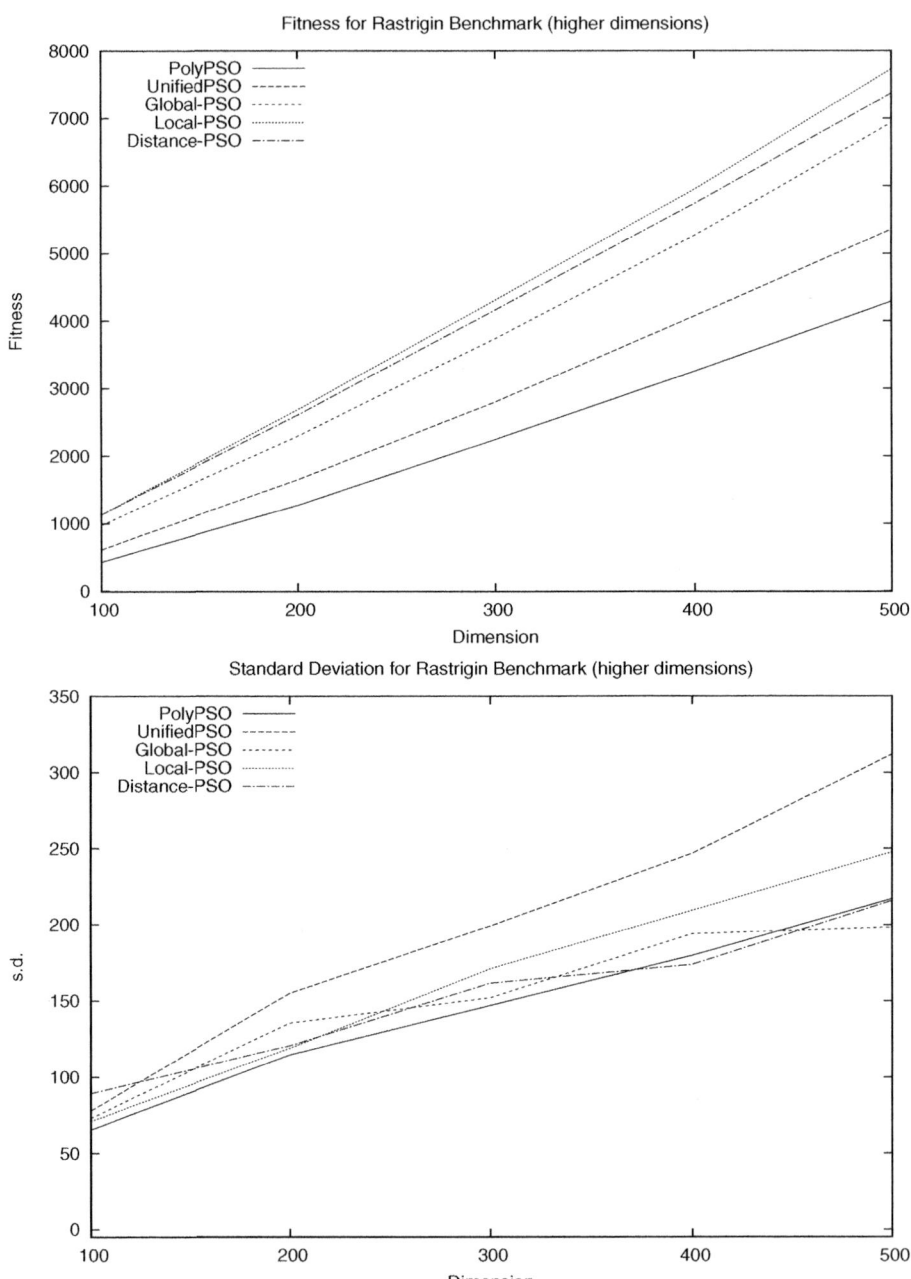

**Fig. 11.** The fitness and standard deviation for the **Rastrigin** Benchmark over all **high** dimensions

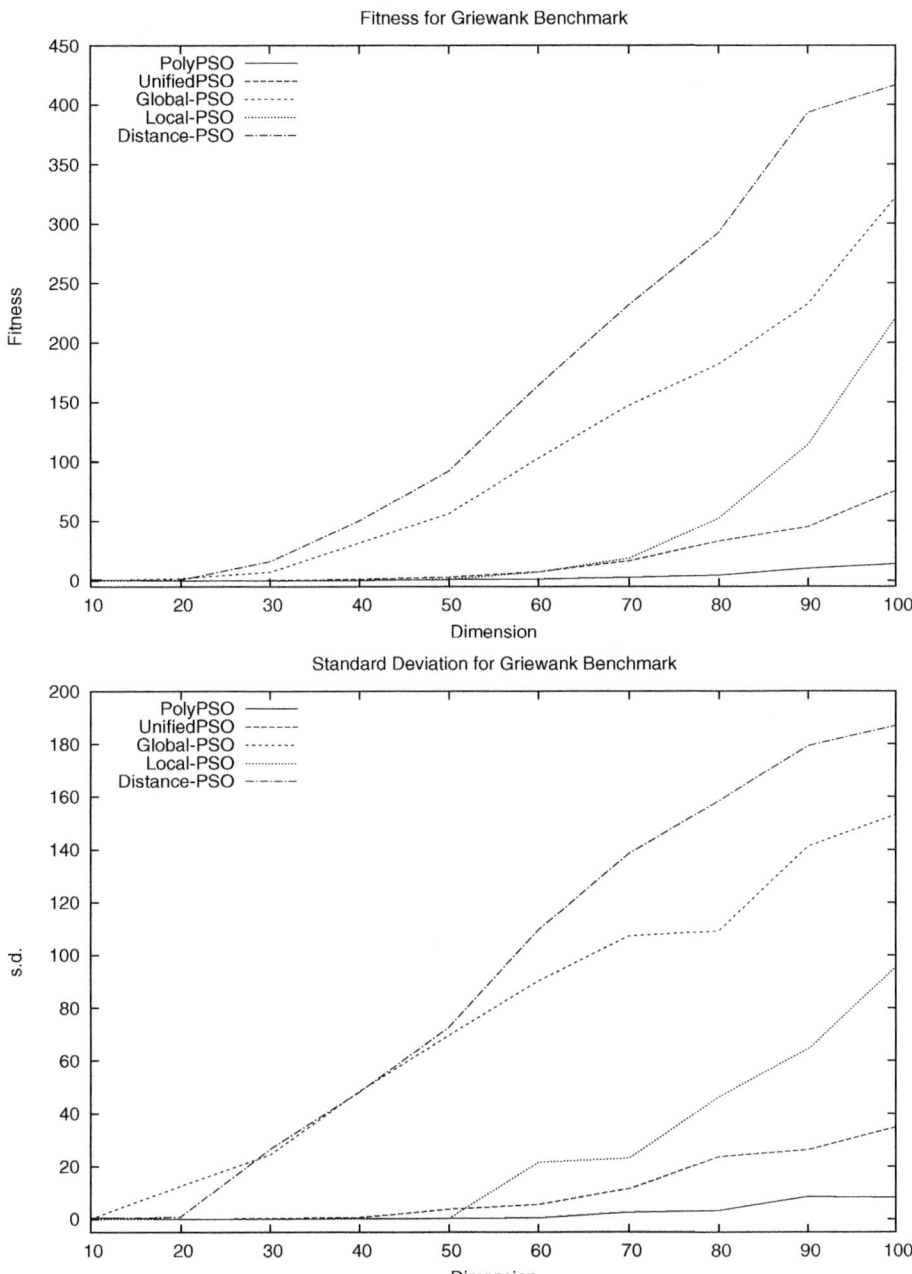

**Fig. 12.** The fitness and standard deviation for the **Griewank** Benchmark over all **low** dimensions

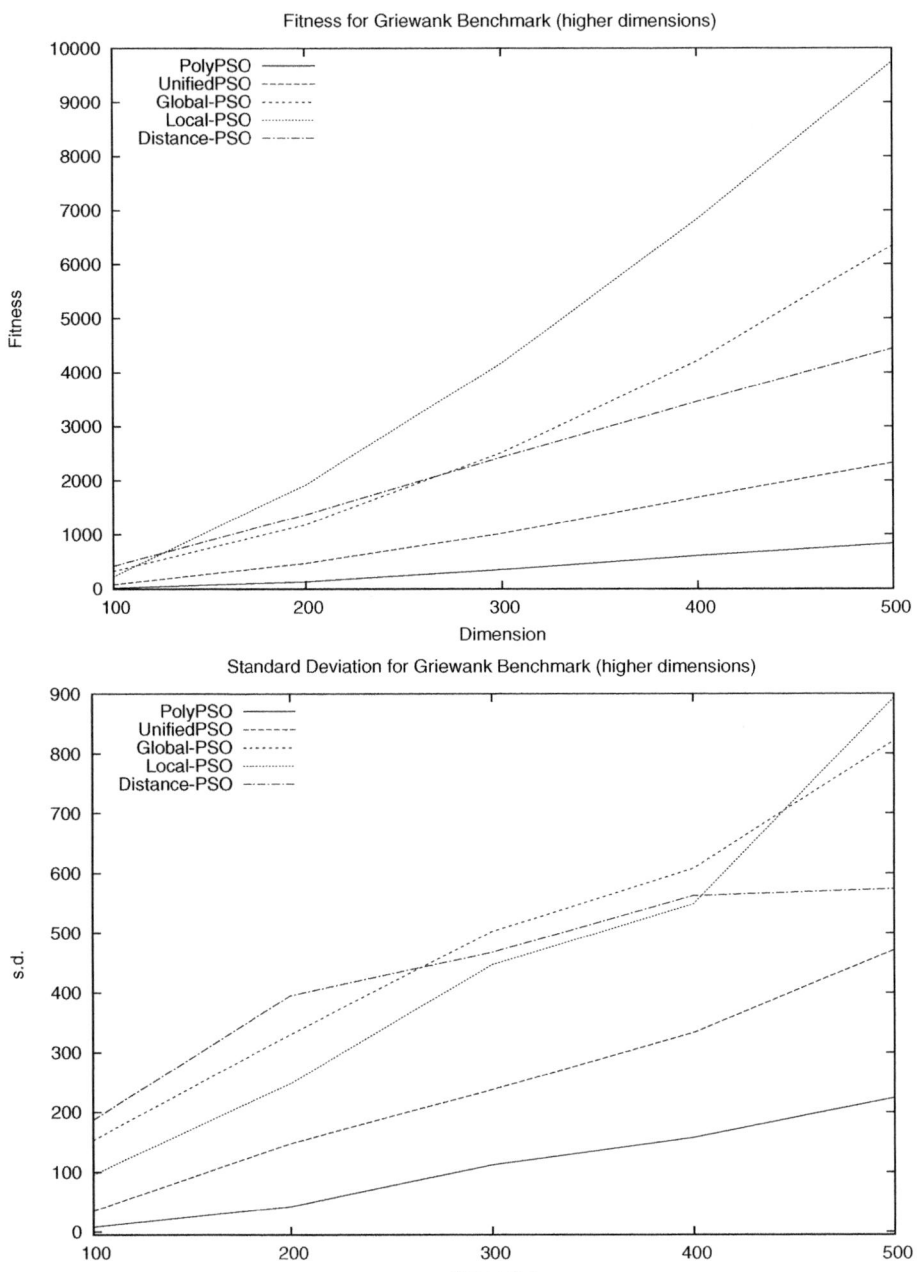

**Fig. 13.** The fitness and standard deviation for the **Griewank** Benchmark over all **high** dimensions

Figures 12 (page 37) and 13 (page 38) show that PolyPSO is the winner and UnifiedPSO the second best over all dimensions. Local-PSO is the third best in lower dimensions, but gets continuously worse in higher dimensions. Contrary to this, the higher the dimension, the better gets Distance-PSO.

The best standard deviation have the PolyPSO, UnifiedPSO and Local-PSO methods (in this order). Merely, approx. at dimension 400 Local-PSO is outperformed by Distance-PSO.

### 6.5 Summary

Considered over all benchmark functions used it can be stated that PolyPSO is the winner in 3 of 4 cases. In one case (Rosenbrock) the UnifiedPSO was slightly better than PolyPSO, which occupies the second rank. In 3 of 4 cases the UnifiedPSO was better than all standard PSO variants. In one case (Sphere) it is only better in higher dimensions. Among the standard PSO methods, Local-PSO is the winner in 3 of 4 cases for lower dimensions. Interestingly, for higher dimensions, Global-PSO performs better in 3 of 4 cases.

## 7   Conclusions

In this paper a new concept called Polymorphic Particle Swarm Optimization was introduced and applied to four typical benchmark functions. Based on a concept of polymorphic equations, the standard PSO update rule was redefined to become a polymorphic update rule. The mathematical expression of this polymorphic update rule is changed on symbolic level based on accumulative histograms and roulette-wheel sampling introducing this way adaptivity. This polymorphic update rule is able to map all 3 standard PSOs known from literature as special cases realizing this way a kind of generalization.

The main aim of this work was to get rid of the topology selection process in PSO. Since PolyPSO performs either as best or second best, it can be used as alternative to the standard PSOs fulfilling this way the main aim.

Furthermore, the higher the dimension, the better PolyPSO outperforms the standard PSO methods. Thus, it could be an interesting candidate for high dimensional problems as for instance in learning hundreds of weights for neural networks or a big number of high dimensional centroids in data clustering.

## References

1. Clerc, M., Kennedy, J.: The particle swarm - explosion, stability, and convergence in a multidimensional complex space. IEEE Transactions on Evolutionary Computation 6(1), 58–73 (2002)
2. Engelbrecht, A.P.: Fundamentals of Computational Swarm Intelligence. Wiley, Chichester (2005) ISBN: 0-470-09191-6
3. Kennedy, J., Eberhart, R.C.: Particle Swarm Optimization. In: IEEE International Conference on Neural Networks, Perth, Australia. IEEE Service Center, Piscataway (1995)

4. Kennedy, J.: Small worlds and mega-minds: Effects of neighborhood topology on particle swarm performance. In: Proceedings of the 1999 Conference on Evolutionary Computation, vol. 3, pp. 1931–1938 (1999)
5. Kennedy, J., Eberhart, R.C.: Swarm Intelligence. Morgan Kaufmann Publishers, San Francisco (2001) ISBN: 1-55860-595-9
6. Kennedy, J., Mendes, R.: Population structure and particle swarm performance. In: Proceedings of the 2002 Congress on Evolutionary Computation, vol. 2, pp. 1671–1676 (2002)
7. Parsopoulos, K.E., Vrahatis, M.N.: Unified particle swarm optimization for tackling operations research problems. In: Proc. IEEE 2005 Swarm Intelligence Symposium, Pasadena (CA), USA, pp. 53–59 (2005)
8. Parsopoulos, K.E., Vrahatis, M.N.: Parameter selection and adaptation in Unified Particle Swarm Optimization. Mathematical and Computer Modelling 46(1-2), 198–213 (2007)
9. Shi, Y.H., Eberhart, R.C.: A Modified Particle Swarm Optimizer. In: IEEE International Conference on Evolutionary Computation, Anchorage, Alaska (1998)
10. Shi, Y., Eberhart, R.C.: Fuzzy adaptive particle swarm optimization. In: Proc. of the 2001 Congress on Evolutionary Computation, vol. 1, pp. 101–106 (2001)
11. Suganthan, P.N.: Particle swarm optimiser with neighbourhood operator. In: Proceedings of the 1999 Congress on Evolutionary Computation, vol. 3, pp. 1958–1962 (1999)
12. Wen, Z., Yutian, L., Clerc, M.: An adaptive PSO algorithm for reactive power optimization. In: Sixth International Conference on Advances in Power System Control, Operation and Management, vol. 1, pp. 302–307 (2003)

# C-Strategy: A Dynamic Adaptive Strategy for the CLONALG Algorithm*

María Cristina Riff[1], Elizabeth Montero[1,2], and Bertrand Neveu[3]

[1] Universidad Técnica Federico Santa María,
Valparaíso, Chile
{Maria-Cristina.Riff,Elizabeth.Montero}@inf.utfsm.cl
[2] Université Nice Sophia Antipolis,
France
[3] Projet COPRIN, INRIA Sophia-Antipolis
France
Bertrand.Neveu@sophia.inria.fr

**Abstract.** The control of parameters during the execution of bio-inspired algorithms is an open research area. In this paper, we propose a new parameter control strategy for the immune algorithm CLONALG. Our approach is based on reinforcement learning ideas. We focus our attention on controlling the number of clones. Our approach provides an efficient and low cost adaptive technique for parameter control. We use instances of the Travelling Salesman Problem. The results obtained are very encouraging.

## 1 Introduction

When we design a bio-inspired algorithm to address a specific problem we need to define a representation, and a set of components to solve the problem. We also need to choose parameter values which we expect will give the algorithm the best results. This process of finding adequate parameter values is a time-consuming task and considerable effort has already gone into automating this process [10]. Researchers have used various methods to find good values for the parameters, as these can affect the performance of the algorithm in a significant way. The best known mechanism to do this is *tuning parameters* on the basis of experimentation, something similar to a generate and test procedure. Taking into account that an immune algorithm run is an intrinsically dynamic adaptive process, we can expect that a dynamic adaptation of the parameters during the search could help to improve the performance of the algorithm, as it has been shown for other metaheuristics [3], [5], [23], [29], [7], [16], [15]. The idea of adapting and self-adapting parameters during the run is not something new, but we need to manage a trade-off between the improvement of the search and the adaptation cost. In this case, the idea is to monitor the search to be able to trigger actions from an adaptive parameter control strategy, in order to improve the performance of the well-known immune algorithm CLONALG [4].

---

* Partially Supported by the Fondecyt Project 1080110.

M.L. Gavrilova et al. (Eds.): Trans. on Comput. Sci. VIII, LNCS 6260, pp. 41–55, 2010.
© Springer-Verlag Berlin Heidelberg 2010

We present a brief revision of the related work in the parameter control domain in the following section. The algorithm CLONALG is briefly described in section 3 with remarks on the aspects of parameter control. We propose a method in section 4 which includes a monitoring task. We have tested the algorithm using our strategy with instances of the Travelling Salesman Problem, that is reported in section 5. The conclusions in section 6 resume our experience from this study and the following trends of this research.

## 2   Related Work

We can classify the parameter selection process into two different methods:*tuning* and *parameter control*, [6]. Tuning, as mentioned above, implies in the worst case a generate and test procedure, in order to define which are the "best" parameter values for a bio-inspired algorithm. Usually, these parameter values are fixed for all the runs of the algorithm. As we mentioned before, Revac [19] has been proposed to tune of evolutionary algorithms in an efficient way. The idea is to use an estimation of distribution algorithms in an iterative process to converge upon a set of good values for the parameters. The ParamILS [12] technique is also available for tuning, it uses a local search strategy to search the parameter values space in an efficient way. The Racing methods [1] use statistical information about the performance of the tuned algorithm to carry out a race between different parameter configurations.

We can expect that the "best" parameter values depend on the step of the algorithm we are using. The main disadvantage of the tuning methods is the time they require to perform the calibration process. For that reason, the idea of designing strategies to allow the algorithm to change its parameter values during its execution appears as a promising way to improve the performance of a bio-inspired algorithm. In this context many ideas have been proposed in the evolutionary research community, including self-adaptability methods. In self-adaptive methods, each individual incorporates information which can influence the parameter values or the evolution of these processes can be used to define some criteria to produce changes, [22], [9], [16].

The parameter control methods can be grouped according to the parameter controlled. For example, we can find strategies which control the population size. The parameter-less algorithm [13] implements a population control performing simultaneous runs of a genetic algorithm with different population sizes, emulating a race among the multiple populations. As the search progresses, parameter-less eliminates small populations which present a worse average performance than larger populations. A recent research of Eiben et al. [7] proposes a strategy to change the population size of the genetic algorithm, taking into account the state of the search. The method is called PRoFIGA. PRoFIGA is able to change the population size according to the convergence and stagnation of the search process.

It is also possible to find strategies which combine methods to control several parameters. In [24], Srinivasa et al. propose the SAMGA method. The SAMGA method incorporates an adaptive parameter control technique to vary the population size, mutation rate and crossover rate of each population of a parallel genetic algorithm. The changes are performed according to the relative performance of one population against the others. Moreover, it dynamically controls the migration rate based on the stagnation of the global search procedure. Recently, Montiel et al. [17] introduced the HEM algorithm which implements an evolutionary algorithm. The HEM algorithm incorporates an adaptive intelligent system which allows the parameter values of the individuals to be controlled in a self-adaptive way. The intelligent system is able to learn about the evolutionary algorithm performance based on expert knowledge and feedback from the search process.

The parameter setting problem, usually suited for evolutionary algorithms, also has been extended to other research areas. In [14] Mezura and Palomeque proposed a parameter control strategy for the constraint-handling mechanism used by a Differential Evolution Algorithm. In their strategy they self-adapted three parameters and one parameter is deterministically controlled. In [28], the authors proposed a two-fold strategy to deterministically control the exploration/exploitation balance during the search and to adaptively coordinate the crossover and mutation operation. This strategy is implemented in an evolutionary multi-objective algorithm with binary representation.

One of the first approaches to parameter control in immune systems was proposed by [8]. In this work Garret proposed a self-adaptive control strategy to control the amount of clones generated from each cell. The amount of clones is established according to predictions of the performance of the algorithm based on the performance that the algorithm showed in the previous iterations. In [11], Hu et al. propose an adaptive control strategy to control the exploration/exploitation rate in the search. In their approach, they divide the repertoire set in three subsets based on their quality. The cells in the best set are mutated according to a micro-mutation process based on the normalized performance of these cells. This approach implements an elitist crossover operator that operates in the two better set of cells. The strategy proposed is able to control the amount of exploitation/exploration according to the variation of two parameters associated to the micro-mutation and crossover operators here defined. The parameter associated to the micro-mutation operator is controlled in an adaptive way and the parameter associated to the crossover operator is controlled in a deterministic way.

The advantage of changing the parameter values during the execution becomes more relevant when we are tackling NP-complete problems. In this paper we introduce a strategy which is able to implement efficient parameter control by using the idea of taking into account the evolution, while also taking action during the execution in order to accelerate the convergence process.

## 3   Description of CLONALG

Artificial Immune Systems (AIS) have been defined as adaptive systems inspired by the immune system and applied to problem solving. In this paper we are interested in CLONALG that is an artificial immune algorithm based on Clonal Selection. CLONALG has successfully been applied to solve various complex

**Standard CLONALG**
**Begin**
Cells = Initialise population of $A$ antibodies                                    (1)
Calculate Fitness (Cells)                                                          (2)
i=1;
**Repeat**
   $S_p$ = Selected **n** best antibodies from Cells               (3)
   $P_c$ = Generate $C$ clones of each antibody from $S_p$         (4)
   $P_c$ = Mutation($P_c$)                                          (5)
   Cells = Cells + $P_c$                                           (6)
   Cells = Selected **n** best antibodies from Cells               (7)
   Cells = Replace worst (Cells, Generate $(A - \mathbf{n})$ New antibodies);   (8)
   $i = i + 1$;
**until** i=Max-number-of-iterations
**End**

**Fig. 1.** Standard version of CLONALG

problems. It follows the basic theory of an immune system response to pathogens. Roughly speaking, the components of the algorithm are cells named antibodies and an antigen that is an invader attacking the immune system. The immune system reproduces those cells that recognise antigens. Cells that match well are cloned (create copies of themselves). The better the match, the more clones. The clones undergo small mutations which may improve their recognition. A selection process retains the best cells as memory cells. The CLONALG pseudocode is shown in figure 1.

The algorithm starts generating a random population of Cells and computing their fitness related to the problem at hand. The iterative process begins by constructing a sub-population $S_p$ of size **n**, composed by the **n** best antibodies belonging the population of Cells. A new population of clones $P_c$ is constructed by generating $C$ clones of each element on $S_p$. This population of clones follows a mutation process, according to a mutation rate, computed using an exponential distribution with parameter $\rho$, in order to improve the evaluation value of the clones. The set Cells is updated including the **n** best antibodies from $P_c$ and incorporating new $(A - n)$ randomly generated antibodies to construct the population of size $A$.

The standard parameters of CLONALG are: $\rho$, $C$, $n$ and $A$. Our study is focused on dynamically controlling parameter $C$. The value of $C$ is related to the exploitation of the search space, because at each iteration the algorithm tries to improve these antibodies applying a mutation procedure.

## 4   Parameter Control Strategy

The key idea in our approach is to design a low computational cost strategy to control the population size of clones in CLONALG. Our aim is to propose and to evaluate a strategy that allows parameter adaptation in an immune based approach according to the problem at hand. We propose an adaptive reinforcement control evaluated in the experimental comparison section.

### 4.1  Adaptive Control

CLONALG works with a set of antibodies as we mentioned before: A population of $C$ clones. The value of this parameter can be controlled in order to guide the trade-off between the intensification and the diversification process of the algorithm. Our idea for this control comes from reinforcement learning. We clearly distinguish between a positive behaviour and a negative one either to reward (increase) or to penalize (reduce) the number of antibodies that belong to the set of clones.

Before we introduce our strategy we require the following definition:

**Definition 1.** *Given a fitness function $F$ to be maximized, a population $P_c$ of $C$ clones and a population $S_p$ of* **n** *selected antibodies. We define a success measure for the parameter $C$ in the $i - th$ iteration, $PS_i(C)$ as:*

$$PS_i(C) = F(BP_c) - F(BS_p) \qquad (1)$$

where $F(BP_c)$ and $F(BS_p)$ are the respective fitness of the best antibody of the populations $P_c$ and $S_p$. The idea is to evaluate if there is an improvement of the best pre-solution found by the algorithm. In this case we reward this situation by increasing the number of $C$ clones, that means more exploitation or intensification of the search f the algorithm. When the algorithm decides to reduce the number of clones it produces that the new population of cells will be completed adding more new cells. From a conceptual point of view it can be interpreted as more exploration or diversification of the search of the algorithm.

Figure 2 shows the structure of the algorithm with the adaptive control. It is important to remark that in lines 3 and 7 the fitness are data available in the original CLONALG code. We explicitly show their evaluations here because we use them in lines 9 and 10. The algorithm computes the $PS_i(C)$ value in order to increase (decrease) $C$ (number of clones) value. Obviously, the algorithm manages extreme situations, as $C$ must be greater than 1.

*Remark 1.* The previous definition and the procedure can easily be adapted when the function to be optimized must be minimized, as we use in the experimental comparison section with the travelling salesman problem

The main motivation of the work reported in this paper is to address the on-line parameter determination problem. We propose a new method which improves the performance of CLONALG and also allows a self-adaptation of the parameter $C$ without adding a significant overhead and without introducing any major changes to its original algorithmic design. The decisions made during the search are based on information available from a monitoring process. Such information allows the algorithm to trigger changes when deemed necessary, based on the potential advantages that such changes could bring. It is also important to keep a proper balance between the level of improvement achieved and the computational cost required to reach it.

**CLONALG-with adaptive control**
**Begin**
Cells = Initialise population of $A$ antibodies                    (1)
$i = 1$;
**Repeat**
  $S_p$ = Selected **n** best antibodies from Cells                   (2)
  $F(BS_p)$= Fitness of the best antibody in $S_p$                   (3)
  $P_c$ = Generate $C$ clones of each antibody from $S_p$            (4)
  $P_c$ = Mutation($P_c$)                               (5)
  $P_c$ = Selected **n** best antibodies from $P_c$                 (6)
  $F(BP_c)$ = Fitness of the best antibody in $P_c$                 (7)
  New-Cells = $P_c$ + Generate $(A - \mathbf{n})$ New antibodies       (8)
  $PS_i(C) = F(BP_c) - F(BS_p)$                           (9)
  **If** $(PS_i(C) > 0)$ **then** $C = C + 1$
    **else If** $C > 2$ **then** $C = C - 1$                     (10)
  $i = i + 1$                                     (11)
  Cells = New-Cells                               (12)
**until** i=Max-number-of-iterations
**End**

**Fig. 2.** Clonal Selection Algorithm with Adaptive Control

# 5   Experimental Comparison

In this section, we experimentally evaluate the CLONALG algorithm using our strategy.

## 5.1   Experimental Set-Up

The hardware platform for the experiments was a PC CPU Intel Core i7-920 4GB RAM, Mandriva 2009.

## 5.2   Travelling Salesman Problem

Travelling Salesman Problem (TSP) is a classical optimization problem, defined as the task of finding the shortest path for visiting $N$ cities and returning to the original point. TSP is a classical combinatorial optimization problem [26], [31], [21]. Several approaches have been proposed to face the TSP using artificial immune systems. The more well-known approach was proposed by de Castro and von Zuben [2] that we briefly present in the following section.

**TSP-CLONALG:** The original version of CLONALG to solve TSP [2] called TSP-CLONALG, has some particularities that differs of its standard version to be noticed. The number of clones of each antibody is calculated according to equation 2. We can observe that the amount of clones is determined by the parameter $\beta$ and the ranking of the antibody $k$. The total amount of clones at each iteration is calculated according to equation 3.

$$C_k = round\left(\frac{\beta \cdot A}{k}\right) \tag{2}$$

$$C = \sum_{k=1}^{n} C_k \tag{3}$$

In order to improve the performance of the standard version of the CLONALG to solve TSP the authors have added two new parameters: The number of randomly generated cells to be included ($d$) after a fixed number of iterations ($iter$). In the original version of CLONALG the random cells were incorporated at each iteration and the number of random cells was fixed at ($A - n$). Thus, this version of the algorithm uses six parameters: $\rho$, $\beta$, $A$, $n$, $d$, $iter$. Using $\beta$ and $A$ it computes the total amount of parameters $C$.

In this paper we compare the performance between TSP-CLONALG using 6 parameters and the standard CLONALG version using our strategy. In our study we have detected that the parameters $\rho$ and $\beta$ are not relevant for the algorithm when it uses our control strategy. Thus, for our approach we just considered 2 parameters ($A$, $n$).

In both versions of CLONALG the representation is a permutation of the cities and the affinity measure is the length of the tour associated to each instantiation or antibody. We have considered as stopping condition when the algorithm is not able to get a better solution during 40 iterations.

Before continuing our description of tests we will describe the tuning method used in our experiments.

## 5.3   Tuning Method: Revac

In our experiments we used the tuning strategy called *Relevance Estimation and Value Calibration* method, Revac. The Revac method was proposed by Eiben & Nannen in [19]. Revac is defined as an estimation of distribution algorithm [20]. It works with a set of parameter configurations, a population. For each parameter, Revac starts the search process with a uniform distribution of values within a given range. As the process advances Revac performs transformation operations with the aim of reducing each parameter distribution to a range of values that performs the best.

Revac method works with a population of 100 parameter configurations and at each iteration generates a new calibration using a crossover and a mutation operator. Revac performs a total amount of 1000 runs of the algorithm to be tuned, that means in average $10^9$ number of evaluations.

## 5.4   Evaluation of the Adaptive Strategy

We report the results using the strategy considering Clones Variations ($C$).

**Scenario 1.** In our experiments we have analyzed the results achieved by de Castro and von Zuben in [2]. In their work they study the results obtained by CLONALG method in an instance of 30 cities from [18] displayed in figure 3. For this scenario we have used the information described in [2] and we have interpreted their *swap* operator as the well-known *2-opt* as the move for mutation. We have used two versions of tuned CLONALG. The first configuration uses the parameter values proposed by the authors of the method, we called this configuration *hand-tuned* configuration. The second one has automatically been obtained by Revac. Table 1 shows the original parameter configuration and this obtained by Revac. In the original version for TSP, the algorithm differs from the classical CLONALG. For TSP two new parameters are included as we mentioned in section 5.2. It considers a set of $d = 60$ random cells to be included after every 20 iterations. Our adaptive technique works on the standard CLONALG, thus these two parameters are not required for the algorithm using control.

**Table 1.** Parameter configurations for the 30 cities instance

| Instance | $\rho$ | $\beta$ | A | **n** | $d$ |
|---|---|---|---|---|---|
| Hand-tuned | 2.5 | 2.0 | 300 | 150 | 60 |
| Revac | 1.2 | 2.8 | 190 | 178 | 97 |

Table 2 shows the performance obtained by the tuned parameter configurations and the adaptive strategy proposed for solving the 30 city instance. The three versions of the algorithm found the optimal value of 48873.

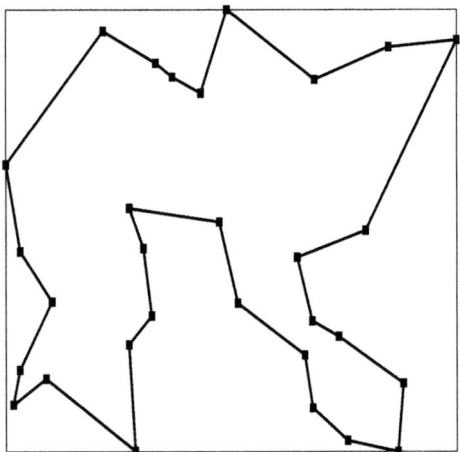

**Fig. 3.** Coordinates of the 30 cities instance

From the results we observe that the C-strategy outperforms all other versions. However it appears using more evaluations than the tuned versions. The algorithm using C-strategy has found the optimal tour 25 times, of the 50 total runs, the original hand-tuned version found 6 times and the version tuned using Revac only found the optimal solution 8 times.

We do not know how many times the algorithm has been run and the corresponding number of evaluations done to obtain the hand-tuned parameters configuration. Revac performed around of $10^8$ number of evaluations in order to obtain the parameter values. These number of evaluations are not included in the Table 2.

**Table 2.** Performance evaluation in the 30 cities instance. AS: Average Solution, $\%\Delta_{Av}$: Relative distance to optimum of the average of 10 runs, $E_{Av}$: Average evaluations, $E_{best}$: Evaluations of the best run, OF: Times the optimum was found.

| Algorithm | AS | $\%\Delta_{Av}$ | $E_{Av}$ | $E_{best}$ | OF |
|---|---|---|---|---|---|
| Hand-tuned | 50795 | 3.9 | 48693 | 51256 | 6 |
| Tuned by Revac* | 50976 | 4.3 | 45649 | 47203 | 8 |
| C-strategy | 50109 | 2.5 | 76413 | 78990 | 25 |

\* Tuned by Revac for 30 cities instance. Revac required around of $10^8$ number of evaluations for tuning this instance. These evaluations are not included in the number of evaluations reported in this table.

**Table 3.** Revac tuning results

| Instance | $\rho$ | $\beta$ | A | n | d |
|---|---|---|---|---|---|
| burma14 | 1.8 | 0.1 | 87 | 87 | 36 |
| ulysses22 | 1.4 | 0.2 | 84 | 84 | 65 |
| eil51 | 1.0 | 0.1 | 120 | 120 | 4 |
| berlin52 | 1.8 | 0.3 | 112 | 70 | 37 |
| st70 | 1.1 | 0.2 | 109 | 20 | 36 |
| eil76 | 1.7 | 0.2 | 142 | 92 | 53 |
| pr76 | 0.8 | 0.1 | 93 | 63 | 28 |
| gr96 | 2 | 0.4 | 133 | 94 | 45 |
| kroA100 | 1.7 | 0.7 | 104 | 95 | 45 |
| eil101 | 2.5 | 0.7 | 147 | 103 | 52 |
| ch130 | 1.4 | 0.8 | 149 | 73 | 52 |
| ch150 | 2.4 | 3.2 | 219 | 164 | 59 |
| gr202 | 3.1 | 0.5 | 152 | 150 | 50 |
| tsp225 | 2.4 | 2.0 | 215 | 136 | 43 |
| a280 | 0.9 | 3.8 | 222 | 203 | 70 |
| pcb442 | 0.7 | 4.1 | 206 | 177 | 59 |
| gr666 | 1.1 | 2.6 | 194 | 174 | 97 |
| pr1002 | 0.5 | 1.4 | 172 | 110 | 66 |

**Scenario 2.** We have tested our technique using symmetric travelling salesman problem instances from TSPLib[1]. The TSPLib is a set of more than 100 instances

---

[1] http://www.iwr.uni-heidelberg.de/groups/comopt/software/TSPLIB95/

of the TSP collected by Professor Gerhard Reinelt in the mid 1990âĂŹs and that
has been extensively used in the metaheuristic research area in the last years [25],
[27], [30].

We have also calibrated the algorithm using Revac to solve other instances
of the TSP. For this, we have also estimated the value of parameter $d$, which
corresponds to the number of random cells to be inserted each 20 iterations. The
results obtained are shown in Table 3. We can observe that, for each instance,
Revac has identified different parameter values. This means that the behaviour
of the algorithm strongly depends on both the problem instance and the values of
its parameters. The amount of selected antibodies, $n$, ranges from around 20% to
95% of the population size and the amount of random antibodies, $d$, incorporated
to the population each 20 generations ranges from 3% to 50% of the population
size. The size of the population of cells $A$ and $\rho$ are also dependent on the
instance at hand. The algorithm using our adaptive strategy does not require to
be re-tuned each time it must solve another new instance of the problem. Thus,
the time invested on tuning is strongly reduced.

Graphs in figure 4 show the relative distance to optimun for each instance
comparing the performance of the C-strategy against to the hand-tuned and
Revac-tuned performance.

**Table 4.** Performance of Hand-tuned configuration. AS: Average Solution, $\%\Delta_{Av}$:
Relative distance to optimum of the average of 10 runs, $E_{Av}$: Average evaluations, BS:
Best solution of 10 runs, $\%\Delta_{Best}$: Relative distance to optimum of the best solution
found, $E_{best}$: Evaluations of the best run.

| Instance | AS | $\%\Delta_{Av}$ | $E_{Av}$ | BS | $\%\Delta_{Best}$ | $E_{best}$ |
|---|---|---|---|---|---|---|
| burma14 | 3333 | 0.3 | 16747 | 3323 | 0.0 | 9824 |
| ulysses22 | 7073 | 0.8 | 36631 | 7013 | 0.0 | 30925 |
| eil51 | 450 | 5.5 | 126175 | 435 | 2.1 | 113980 |
| berlin52 | 8093 | 7.3 | 135670 | 7787 | 3.2 | 135263 |
| st70 | 720 | 6.6 | 210339 | 693 | 2.7 | 213953 |
| eil76 | 589 | 9.4 | 207454 | 559 | 3.9 | 239400 |
| pr76 | 115678 | 7.0 | 244200 | 112825 | 4.3 | 291478 |
| gr96 | 59110 | 7.1 | 343631 | 57276 | 3.7 | 346040 |
| kroA100 | 22874 | 7.5 | 374095 | 21596 | 1.5 | 407959 |
| eil101 | 685 | 8.9 | 315624 | 672 | 6.8 | 324708 |
| ch130 | 6661 | 9.0 | 510299 | 6448 | 5.5 | 519051 |
| ch150 | 7248 | 11.0 | 577297 | 6906 | 5.8 | 603625 |
| gr202 | 43353 | 8.0 | 884525 | 42296 | 5.3 | 820052 |
| tsp225 | 4323 | 10.4 | 984360 | 4234 | 8.1 | 1069647 |
| a280 | 2965 | 15.0 | 1338809 | 2923 | 13.3 | 1356322 |
| pcb442 | 58021 | 14.3 | 2328181 | 56343 | 11.0 | 2385774 |
| gr666 | 338838 | 15.1 | 4014526 | 332335 | 12.9 | 4210456 |
| pr1002 | 308884 | 19.2 | 6568289 | 299594 | 15.7 | 6749630 |

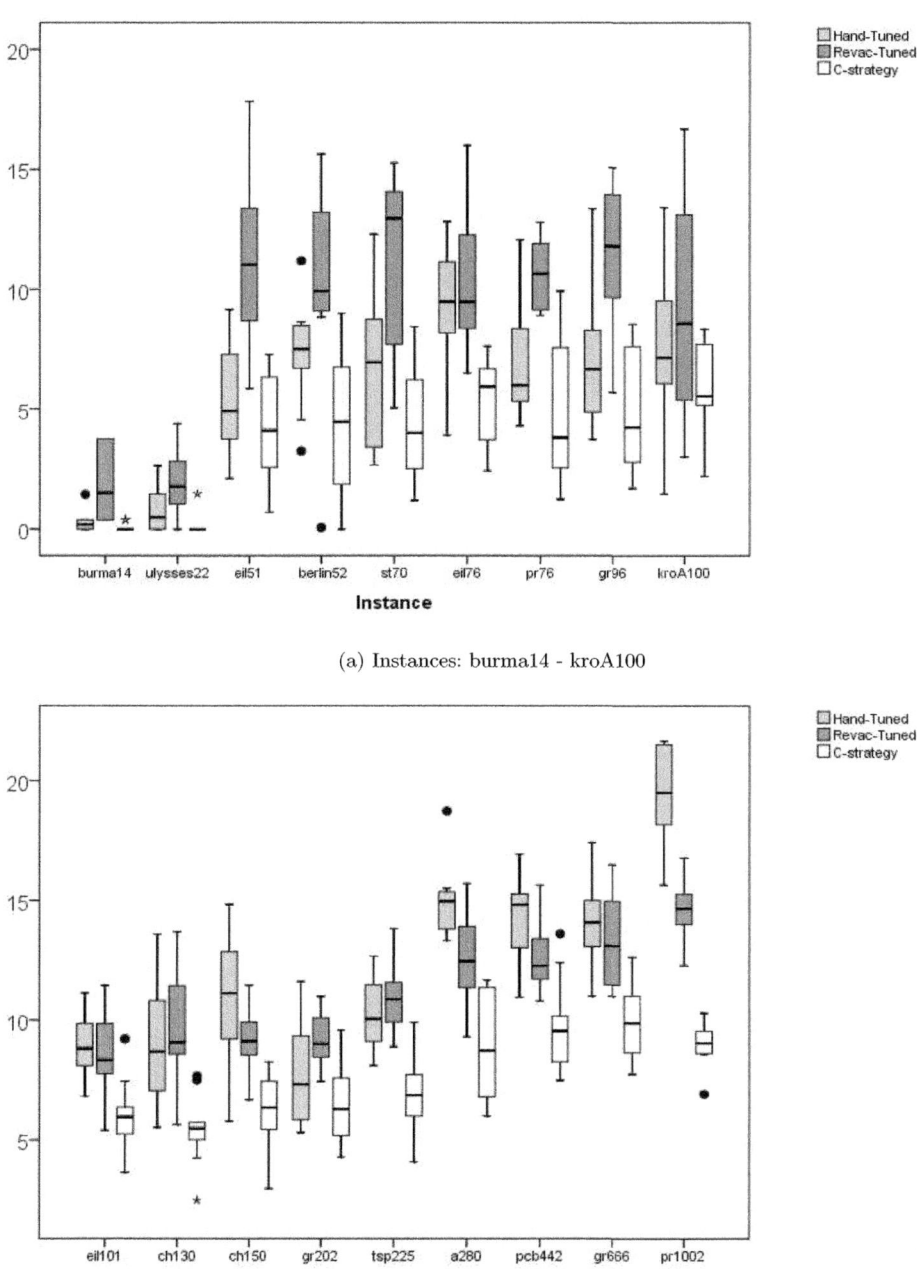

(a) Instances: burma14 - kroA100

(b) Instances: eil101 - pr1002

**Fig. 4.** Performance: Hand-Tuned, Revac-Tuned and C-strategy

**Table 5.** Performance of Revac-tuned configurations. AS: Average Solution, $\%\Delta_{Av}$: Relative distance to optimum of the average of 10 runs, $E_{Av}$: Average evaluations, BS: Best solution of 10 runs, $\%\Delta_{Best}$: Relative distance to optimum of the best solution found, $E_{best}$: Evaluations of the best run, $\Delta^{+}$: Improvement of relative distance to optimum compared to the hand-tuned version.

| Instance | AS | $\%\Delta_{Av}$ | $E_{Av}$† | BS | $\%\Delta_{Best}$ | $E_{best}$† | $\Delta^{+}$ |
|---|---|---|---|---|---|---|---|
| burma14 | 3388 | 2.0 | 518 | 3336 | 0.4 | 707 | -1.7 |
| ulysses22 | 7150 | 2.0 | 1561 | 7013 | 0.0 | 1823 | -1.1 |
| eil51 | 473 | 11.1 | 4052 | 451 | 5.9 | 4256 | -5.5 |
| berlin52 | 8305 | 10.1 | 9799 | 7547 | 0.1 | 9565 | -2.8 |
| st70 | 750 | 11.1 | 13008 | 709 | 5.0 | 17287 | -4.5 |
| eil76 | 594 | 10.4 | 12049 | 573 | 6.5 | 13426 | -1.0 |
| pr76 | 119710 | 10.7 | 7854 | 117792 | 8.9 | 6036 | -3.7 |
| gr96 | 61503 | 11.4 | 33120 | 58362 | 5.7 | 39705 | -4.3 |
| kroA100 | 23285 | 9.4 | 56371 | 21924 | 3.0 | 50938 | -1.9 |
| eil101 | 683 | 8.6 | 122179 | 663 | 5.4 | 144320 | 0.2 |
| ch130 | 6708 | 9.8 | 124983 | 6455 | 5.6 | 131313 | -0.8 |
| ch150 | 7124 | 9.1 | 724714 | 6964 | 6.7 | 749362 | 1.9 |
| gr202 | 43833 | 9.1 | 274549 | 43150 | 7.4 | 306989 | -1.2 |
| tsp225 | 4343 | 10.9 | 849222 | 4264 | 8.9 | 819274 | -0.5 |
| a280 | 2905 | 12.6 | 2227123 | 2819 | 9.3 | 2303958 | 2.3 |
| pcb442 | 57221 | 12.7 | 3372096 | 56269 | 10.8 | 3344210 | 1.6 |
| gr666 | 333980 | 13.5 | 3779875 | 327663 | 11.3 | 3979137 | 1.7 |
| pr1002 | 297017 | 14.7 | 4857562 | 290840 | 12.3 | 4965511 | 4.6 |

† The amount of evaluations specified does not include the around of $10^{9}$ evaluations performed by the tuning process.

Our strategy helps the algorithm to improve the quality of the solutions found by the tuned versions. The greater is the number of cities the more noteworthy are the results obtained by the C-strategy.

## 6    Conclusions

We propose in this paper a strategy for an adaptive parameter control for CLON-ALG.

For intensification, our strategy works with the number of clones to be improved. When the improvement procedure has been applied successfully the algorithm increases the number of clones which will follow the improvement process in the next iteration.

In our design we have taken into account the overhead produced by including our strategy. It is a low cost strategy. The main advantage of using adaptive strategies to improve the behavior of the CLONALG comes from non-existing pre-execution time. The tuning techniques can obtain very good results to find tuned parameters for a given problem, but the parameter configuration step

**Table 6.** Performance of C-strategy. AS: Average Solution, $\%\Delta_{Av}$: Relative distance to optimum of the average of 10 runs, $E_{Av}$: Average evaluations, BS: Best solution of 10 runs, $\%\Delta_{Best}$: Relative distance to optimum of the best solution found, $E_{best}$: Evaluations of the best run, $\Delta^+$: Improvement of relative distance to optimum compared to the hand-tuned version.

| Instance | AS | $\%\Delta_{Av}$ | $E_{Av}$ | BS | $\%\Delta_{Best}$ | $E_{best}$ | $\Delta^+$ |
|---|---|---|---|---|---|---|---|
| burma14 | 3326 | 0.1 | 13326 | 3323 | 0.0 | 13069 | 0.2 |
| ulysses22 | 7023 | 0.1 | 37915 | 7013 | 0.0 | 39049 | 0.7 |
| eil51 | 444 | 4.2 | 280361 | 429 | 0.7 | 212817 | 1.3 |
| berlin52 | 7870 | 4.4 | 325300 | 7542 | 0.0 | 276394 | 2.9 |
| st70 | 705 | 4.5 | 624727 | 683 | 1.2 | 673642 | 2.1 |
| eil76 | 567 | 5.3 | 715914 | 551 | 2.4 | 649833 | 4.1 |
| pr76 | 113273 | 4.7 | 742192 | 109504 | 1.2 | 838858 | 2.2 |
| gr96 | 57878 | 4.8 | 1484345 | 56145 | 1.7 | 1653666 | 2.2 |
| kroA100 | 22545 | 5.9 | 1643365 | 21753 | 2.2 | 1629237 | 1.5 |
| eil101 | 667 | 6.0 | 1394766 | 652 | 3.7 | 1300641 | 2.9 |
| ch130 | 6444 | 5.5 | 2803527 | 6263 | 2.5 | 3069959 | 3.5 |
| ch150 | 6931 | 6.2 | 3969191 | 6722 | 3.0 | 4137293 | 4.9 |
| gr202 | 42771 | 6.5 | 7422549 | 41886 | 4.3 | 41886 | 1.4 |
| tsp225 | 4186 | 6.9 | 9877811 | 4076 | 4.1 | 9599825 | 3.5 |
| a280 | 2809 | 8.9 | 17376294 | 2734 | 6.0 | 17240650 | 6.1 |
| pcb442 | 55693 | 9.7 | 45624518 | 54588 | 7.5 | 49324914 | 4.6 |
| gr666 | 323210 | 9.8 | 117452296 | 317200 | 7.8 | 120874559 | 5.3 |
| pr1002 | 282506 | 9.1 | 286994425 | 276963 | 6.9 | 280607399 | 10.2 |

requires a huge amount of time compared to the time needed for the tuned algorithm to find a solution.

The original TSP-CLONALG algorithm uses 6 parameters. Accorditong to the results obtained using Revac, the value of these parameters are different for different instances of the TSP. Thanks to our dynamic control strategy the number of parameters to solve TSP is reduced to only 2. Moreover, these 2 parameters require only one tuning step whatever is the TSP instance to be solved from these used in our tests.

The algorithm using our adaptive strategy does not require to be re-tuned for solving new instances of the problem. Thus, the time invested for tuning is strongly reduced. The strategy that is proposed helps CLONALG to improve its performance when it is solving Travelling Salesman Problem instances. The C-strategy is able to improve the performance in all instances tested with a relative average improvement of 3.3.

As future work we would like to evaluate our strategy in CLONALG using other hard combinatorial problems. We are also considering to include this strategy in other kinds of artificial immune algorithms based on Immune Network Models. We are working to include this strategy in CD-NAIS, an immune algorithm that solves hard instances of Constraint Satisfaction Problems. A promising research area is the collaboration between various parameter control strategies.

# References

1. Birattari, M., Stützle, T., Paquete, L., Varrentrapp, K.: A racing algorithm for configuring metaheuristics. In: Proceedings of the Genetic and Evolutionary Computation Conference, New York-United States, July 2002, pp. 11–18. Morgan Kaufmann, San Francisco (2002)
2. De Castro, L.N., Von Zuben, F.: The clonal selection algorithm with engineering applications. In: Proceedings of Workshop on Artificial Immune Systems and their Apllications, GECCO, pp. 36–37. Morgan Kaufmann, San Francisco (2000)
3. Davis, L.: Adapting operator probabilities in genetic algorithms. In: Proceedings of the third international conference on Genetic algorithms, pp. 61–69. Morgan Kaufmann, San Francisco (1989)
4. de Castro, L.N., Timmis, J.: Artificial Immune Systems: A New Computational Intelligence Approach. Springer, Heidelberg (2002)
5. Deb, K., Agrawal, S.: Understanding interactions among genetic algorithm parameters. In: Foundations of Genetic Algorithms, vol. 5, pp. 265–286. Morgan Kaufmann, San Francisco (1999)
6. Eiben, A.E., Hinterding, R., Michalewicz, Z.: Parameter control in evolutionary algorithms. IEEE Transactions on Evolutionary Computation 3, 124–141 (1999)
7. Eiben, A.E., Marchiori, E., Valkó, V.A.: Evolutionary algorithms with on-the-fly population size adjustment. In: Yao, X., Burke, E.K., Lozano, J.A., Smith, J., Merelo-Guervós, J.J., Bullinaria, J.A., Rowe, J.E., Tiño, P., Kabán, A., Schwefel, H.-P. (eds.) PPSN 2004. LNCS, vol. 3242, pp. 41–50. Springer, Heidelberg (2004)
8. Garrett, S.M.: Parameter-free, adaptive clonal selection. In: IEEE Congress on Evolutionary Computation, vol. 1, pp. 1052–1058 (2004)
9. Gómez, J.: Self adaptation of operador rates in evolutionary algorithms. In: Deb, K., et al. (eds.) GECCO 2004. LNCS, vol. 3102, pp. 1162–1173. Springer, Heidelberg (2004)
10. Hinterding, R., Michalewicz, Z., Eiben, A.E.: Adaptation in evolutionary computation: A survey. In: IEEE International Conference on Evolutionary Computation, pp. 65–69 (1997)
11. Hu, J., Guo, C., Li, T., Yin, J.: Adaptive clonal selection with elitism-guided crossover for function optimization. In: International Conference on Innovative Computing, Information and Control, pp. 206–209 (2006)
12. Hutter, F., Hoos, H., Stützle, T.: Automatic algorithm configuration based on local search. In: Proceedings of the Twenty-Second Conference on Artifical Intelligence, pp. 1152–1157 (2007)
13. Lobo, F.G., Goldberg, D.E.: The parameter-less genetic algorithm in practice. Information Sciences 167(1-4), 217–232 (2004)
14. Mezura-Montes, E., Palomeque-Ortiz, A.G.: Parameter control in differential evolution for constrained optimization. In: IEEE International Conference on E-Commerce Technology, pp. 1375–1382 (2009)
15. Montero, E., Riff, M.C., Basterrica, D.: Improving MMAS using parameter control. In: IEEE Congress on Evolutionary Computation, Hong-Kong, June 2008, pp. 4007–4011 (2008)
16. Montero, E., Riff, M.C.: Self-calibrating strategies for evolutionary approaches that solve constrained combinatorial problems. In: An, A., Matwin, S., Raś, Z.W., Ślęzak, D. (eds.) Foundations of Intelligent Systems. LNCS (LNAI), vol. 4994, pp. 262–267. Springer, Heidelberg (2008)

17. Montiel, O., Castillo, O., Melin, P., Díaz, A.R., Sepúlveda, R.: Human evolutionary model: A new approach to optimization. Information Sciences 177(10), 2075–2098 (2007)
18. Moscato, P., Fontanari, J.F.: Stochastic versus deterministic update in simulated annealing. Physics Letters A 146(4), 204–208 (1990)
19. Nannen, V., Eiben, A.E.: Relevance estimation and value calibration of evolutionary algorithm parameters. In: Joint International Conference for Artificial Intelligence (IJCAI), pp. 975–980 (2006)
20. Pelikan, M., Goldberg, D.E., Lobo, F.G.: A Survey of Optimization by Building and Using Probabilistic Models. Computational Optimization and Applications 21(1), 5–20 (2002)
21. Richter, D., Goldengorinand, B., Jäger, G., Molitor, P.: Improving the efficiency of helsgauns lin-kernighan heuristic for the symmetric tsp. In: Proceedings of the Fourth Workshop on Combinatorial and Algorithmic Aspects of Networking, pp. 99–111 (2007)
22. Riff, M.C., Bonnaire, X.: Inheriting parents operators: a new dynamic strategy to improve evolutionary algorithms. In: Hacid, M.-S., Raś, Z.W., Zighed, D.A., Kodratoff, Y. (eds.) ISMIS 2002. LNCS (LNAI), vol. 2366, pp. 333–341. Springer, Heidelberg (2002)
23. Smith, J.E., Fogarty, T.C.: Operator and parameter adaptation in genetic algorithms. Soft Computing - A Fusion of Foundations, Methodologies and Applications 1(2), 81–87 (1997)
24. Srinivasa, K.G., Venugopal, K.R., Patnaik, L.M.: A self-adaptive migration model genetic algorithm for data mining applications. Information Sciences 177(20), 4295–4313 (2007)
25. Stützle, T., Hoos, H.: Max-min ant system and local search for the traveling salesman problem. In: IEEE International Conference on Evolutionary Computation, pp. 309–314 (1997)
26. Stützle, T., Grün, A., Linke, S., Rüttger, M.: A comparison of nature inspired heuristics on the traveling salesman problem. In: Proceedings of the Parallel Problem Solving from Nature (PPSN VI), pp. 661–670. Springer, Heidelberg (2000)
27. Sun, W.-D., Xu, X.-S., Dai, H.-W., Tang, Z., Tamura, H.: An immune optimization algorithm for tsp problem. In: SICE 2004 Annual Conference, vol. 1, pp. 710–715 (2004)
28. Tan, K.C., Chiam, S.C., Mamun, A.A., Goh, C.K.: Balancing exploration and exploitation with adaptive variation for evolutionary multi-objective optimization. European Journal of Operational Research 197(2), 701–713 (2009)
29. Tuson, A., Ross, P.: Adapting operator settings in genetic algorithms. Evolutionary Computation 6(2), 161–184 (1998)
30. Yang, J., Wu, C., Pueh Lee, H., Liang, Y.: Solving traveling salesman problems using generalized chromosome genetic algorithm. Progress in Natural Science 18(7), 887–892 (2008)
31. Zhang, W., Looks, M.: A novel local search algorithm for the traveling salesman problem that exploits backbones. In: Proceedings of the International Joint Conferences on Artificial Intelligence (IJCAI 2005), pp. 343–350 (2005)

# A Comparison of Genotype Representations to Acquire Stock Trading Strategy Using Genetic Algorithms*

Kazuhiro Matsui and Haruo Sato

Department of Computer Science, College of Engineering, Nihon University
1 Nakagawara, Tokusada, Tamura-machi, Koriyama, 963-8642, Japan
Tel.: +81-24-956-8829;
Fax: +81-24-956-8863
matsui@cs.ce.nihon-u.ac.jp, sato@cs.ce.nihon-u.ac.jp

**Abstract.** Automatic trading methods, such as algorithmic trading, are important issues in recent financial markets. Various approaches have been proposed in this context. We compare some genotype coding methods of technical indicators and their parameters to acquire stock trading strategy using genetic algorithms (GAs) in this paper. In previous related works, the locus-based representation was widely employed for encoding technical indicators on chromosomes in GAs, and the direct coding was also widely adopted for encoding the parameters of the indicators. However, we show that these conventional methods are not so effective for the GA search. Therefore, we propose a new genotype coding methods, namely the allele-based indirect representation. We examine the performance of the proposed and conventional coding methods in stock trading for twenty companies in the first section of the Tokyo Stock Exchange for recent ten years. In our empirical results, the allele-based indirect representation is superior to the other ones both on the cumulative profits and the computational costs.

**Keywords:** Genetic algorithm, Genotype representation, Automatic stock trading.

## 1 Introduction

Automatic trading methods, such as algorithmic trading, are expanding rapidly in recent financial markets. Various works have been proposed in applications of computational intelligence methodologies in finance [1]. In these methodologies, evolutionary computation, such as the genetic algorithm (GA)[2], is promising because of their robustness, flexibility and powerful ability for search.

Several works have been done for acquiring trading strategy using evolutionary computation [3], [4], [5], and [6]. Their methods are based on technical analysis,

* This work was partially supported by the Grant-in-Aid for Scientific Research (C) 20500215, Japan Society for the Promotion of Science.

M.L. Gavrilova et al. (Eds.): Trans. on Comput. Sci. VIII, LNCS 6260, pp. 56–70, 2010.

which is one of the two basic types of approaches in stock trading. Technical analysis is an attempt to forecast the future direction of prices by analyzing past market data, such as price and volume. The other type of the approaches to analyze stock trading is fundamental analysis, which focuses on analyzing financial statements and management. The above works to acquire trading strategy adopted technical analysis because it was easily applied to automatic trading in comparing with fundamental analysis.

Various kinds of genotype-phenotype coding are applied in these works. However, it is not clear which representation is better than others. In this paper, we compare some genotype representations in terms of coding for technical indicators and their parameters. In conventional coding methods for technical indicators, a locus-based representation has been widely used. This representation causes chromosomes in GAs to be too long when many technical indicators are used. The conventional coding methods also employed simple binary chromosomes for parameters of technical indicators. However, the efficiency of the binary coding is relatively low for searching spaces of parameters. Therefore, we propose a new genotype representation to solve these problems. Our representation is called the allele-based indirect representation. We compare it with some conventional methods and show the effectiveness of our method.

This paper has two objectives. The first is to compare genotype representations for acquiring stock trading strategy. We compare them in two aspects: the locus-based representation versus the allele-based one, and the direct coding versus the indirect one. The second objective is to show the superiority of our method to the other representations.

This paper is organized as follows: In Section 2, we summarize the concept of technical analysis in stock trading. We describe the details of genotype representation in Section 3. Section 4 contains our trading method and the empirical results are followed in Section 5. Sections 6 and 7 are discussion and conclusions respectively.

## 2   Technical Analysis in Stock Trading

We have two basic approaches to analyze markets: fundamental analysis and technical one. The former is based on analyzing financial statements, management, and competitive advantages of companies. The latter is based on the past patterns of changes of share prices. In technical analysis, many indicators are used for trading. They are calculated from past share prices and/or volumes. Generally, technical indicators have some parameters. For example, moving average has a parameter, namely *period*, which is used as the denominator of the averaging calculation. Various derived indicators, such as *10-days moving average*, *25-days moving average, etc.*, can be defined with the period parameter. In this paper, we use many technical indicators and their associated parameters for automatic trading.

Various types of technical indicators are known in traders, but they have two problems. First, it is difficult to select suitable indicators for trading. Second, it is also hard to determine optimal parameters for the selected indicators. In this paper, we apply GAs to these problems. Both technical indicators and their associated parameters are encoded on chromosomes of individuals in GAs and we apply the genetic search to acquire effective combinations of technical indicators and their parameters for trading. The aims of this paper are to compare various methods of genotype representations and to clarify the effectiveness of our new method, which is described in the next section.

## 3    Genotype Representation

### 3.1    Related Works

Several related works have been done on automatic trading strategy using evolutionary computation methodologies. A method to search effective combinations of technical indicators using genetic algorithms was proposed [3]. This method is similar to the locus-based representation described in Section 3.3. The genetic search of this method was limited to technical indicators. Their associated parameters were fixed through the search.

On the other hand, some methods to search the optimal parameters to evaluate technical indicators using binary-coded genetic algorithms were proposed [5], [6]. The searches of these methods were limited to only the parameters for several pre-determined indicators. They were fixed through the genetic search. These genotype representations correspond to the direct coding described in the following subsection, and we compare them with our proposed methods in our experiments described in Section 5.

### 3.2    Parameter Encoding

It is necessary to encode parameters on chromosomes of individuals for genetic search of trading strategy. In the related works, such as [5] and [6], binary coded GAs are mainly used for this aim. However, they involve two shortcomings. First, binary coded GAs generally divide their search-ranges at *regular* intervals and assign each value to each binary code. However, it is often not desirable to divide the range at regular intervals whose width are equal to each other. For examples, suppose that binary coded GAs search the period of the moving average of prices, as shown in Figure 1. In comparing among shorter periods, such as four and five days, it may causes different profits in short-term trading. On the other hand, in comparing among longer periods, such as 99 and 100 days, it will be expected that the difference of profits is little although the difference of these periods is same as one day. Thus, the simple binary coding to divide the search range at regular intervals is not always suitable for the search of parameters. We refer the conventional method of binary coding as *the direct coding* in the following sections.

The second shortcoming of binary coding is the aptitude for the search of real values. In several researches of real-coded GAs (for examples, [7] and [8]),

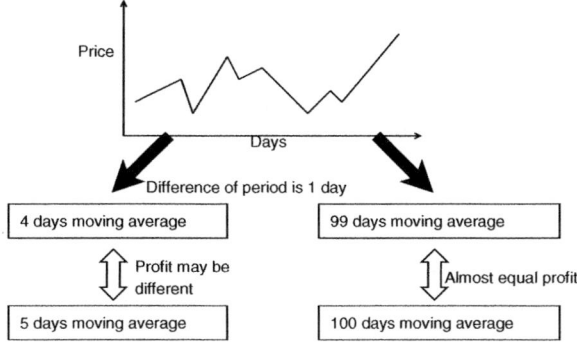

**Fig. 1.** A shortcoming of binary-coded GA

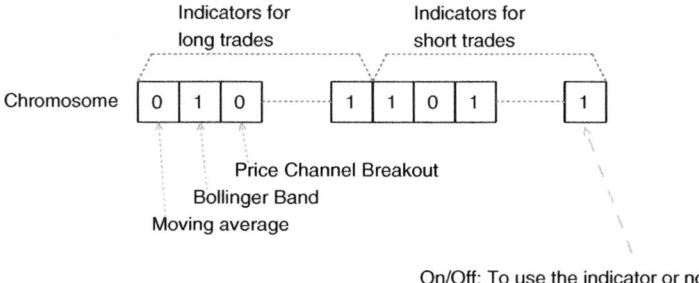

**Fig. 2.** General concept of the locus-based representation

it turned out that binary-coded GAs are not appropriate for the search of real values. Since technical indicators often have real parameters for the evaluation, it is not desirable to use binary-coded GAs for the search of the parameters.

In this paper, we propose a new coding method of parameters. Generally, there often are typical values which are widely used to evaluate technical indicators. Therefore, we restrict the ranges of the genetic search to the set of these typical values. For examples, in the above case of the moving average of prices, we are able to restrict the range to a set of values, such as $\{5, 10, 20, 50, 100\}$. This method makes the space of the genetic search to be smaller than the conventional direct coding. Thus, the efficiency of the genetic search is expected to be higher than the conventional one, and the computational cost will be extremely reduced. We refer the new method of coding as *the indirect coding* in this paper.

### 3.3 Locus-Based Representation

In this paper, we compare two types of the genotype representation of technical indicators. The first is the locus-based representation. Its general concept is shown in Figure 2. This representation assigns each technical indicator to each

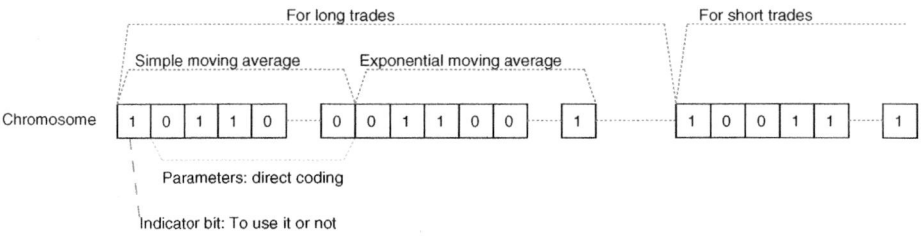

**Fig. 3.** Locus-based direct representation

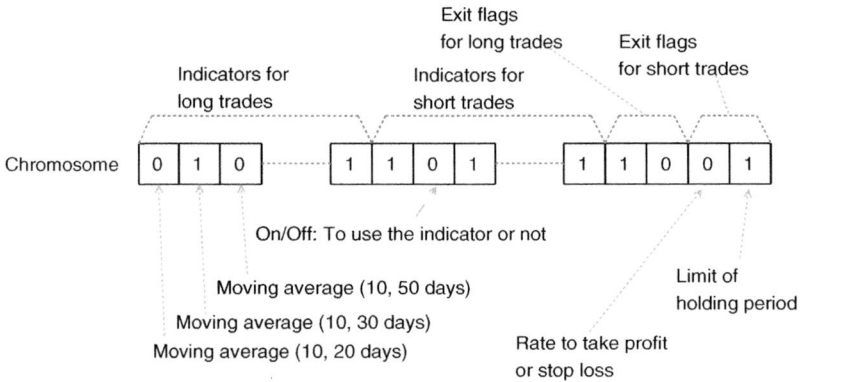

**Fig. 4.** Locus-based indirect representation

locus, which is a bit position on chromosomes. For example, the first bit is assigned to the indicator *"moving average"* in Figure 2. The indicator is used for trading when the assigned bit is "1," and it is not used in the opposite case.

We compare two subtypes of this representation in this paper: the locus-based direct representation and the locus-based indirect representation. In the former, the parameters are represented in the direct coding. Each indicator and its parameters are divided on the chromosomes, as shown in Figure 3. This concept is similar to the conventional representation proposed in [6].

The latter assigns a technical identifier (ID) to each locus, as shown in Figure 4. The ID is a combination of a technical indicator and its associated parameters which are represented in the indirect coding. For example, the first bit is assigned to the ID *"the crossover of moving average between ten and twenty days"* in Figure 4. Note that technical indicators and their parameters are combined on single bits and the parameters are encoded in the indirect coding.

Four additional bits are appended to the end of chromosomes. These bits are the exit flags which consist of two sections. The first section is the first two bits which are used for long trades. The second is the last two bits for short trades. Both of the sections are composed of two bits. The first bit is the exit flag on the rate to take profit or stop loss. When this flag is "on," the system takes

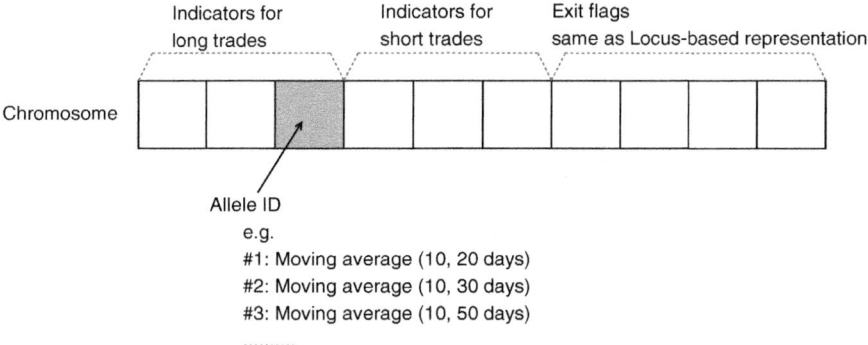

**Fig. 5.** Allele-based indirect representation

profit or stops loss for the trade if the profit/loss reaches the predetermined rate. The second bit is the flag on the limit of the holding period of the trade. When this flag is "on," the system exits the position if its holding period reaches the predetermined limit. We can apply ordinary binary-coded GAs easily to the locus-based representation because the chromosomes are coded in binary strings.

The chromosome length excluding the exit flags in the locus-based indirect representation is equal to the total number of candidates of the IDs. Thus, the length becomes long when the number of the candidates increase.

Note that the locus-based *direct* representation involves the two shortcomings of binary-coded GAs described in the previous subsection. However, the locus-based *indirect* representation is independent from the shortcomings because the indirect coding restricts the search space to the set of typical representative values for the parameters.

### 3.4   Allele-Based Representation

The second type of genotype coding is the allele-based representation. Figure 5 shows the concept of the indirect version of this representation. The allele, which is a value on a locus, takes various IDs which represent technical indicators and their combined parameters in the indirect coding. For example, Allele #1 is assigned to the indicator *"the crossover of moving average between ten and twenty days"* in Figure 5.

In the allele based coding, the four exit flags are also added to the end of the chromosomes.

Generally, the allele-based representation makes the length of chromosome to be much shorter than the locus-based representation. The total number of alleles is identical to the total number of candidates of IDs. Note that it is necessary to determine the length of chromosomes in advance.

## 4   Methods

### 4.1   Overview

We show the overview of our system in Figure 6. The flow of our method is the following:

1. Prepare a set of historic share prices and divide it to the training dataset and the test one.
2. Determine candidates of technical indicators and their parameters.
3. Apply the GA to find effective indicators and their parameters for stock trading on the training dataset.
4. Run the simulation of automatic trading for the test dataset with the indicators and their parameters which have been found by the previous genetic search.

The aim of this process is to maximize the profit on the test dataset, not on the training dataset. Thus, it is important to keep off overfitting on the training dataset.

Note that the GA is applied only on the training phase (Step 3), not on the test phase (Step 4).

The trading rules adopted in our system are applied *daily*. When a trading rule is matched in a day, the system opens or closes the position *at the opening price on the next day.*

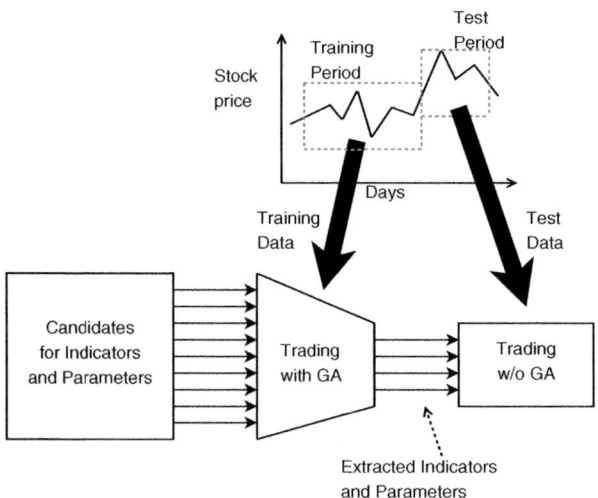

**Fig. 6.** Overview of our system

## 4.2   Technical Indicators

We use the following technical indicators in this paper:

1. Simple Moving Average Crossover (SMA)
   The SMA is a simple average of closing prices for the last $n$ days. We define the SMA as follows:

$$\mathrm{SMA}_n(t) = \frac{1}{n} \sum_{i=0}^{n-1} c_{t-i}, \tag{1}$$

   where $c_t$ is the closing price at the day $t$, $n$ is the parameter which determines the period to calculate the SMA.
   In our experiments, we apply the following rules: For a long trade, enter when a shorter-period SMA crosses a longer-period SMA and exit when the opposite occurs. A short trade is the contrary of the long one.

2. Exponential Moving Average Crossover (EMA)
   The EMA is an exponentially weighted average of closing prices for the last $n$ days, as follows:

$$\mathrm{EMA}_n(t) = \begin{cases} \mathrm{SMA}_n(t) & (t = 0) \\ \mathrm{EMA}_n(t-1) & \\ \quad + \alpha\,(c_t - \mathrm{EMA}_n(t-1)) & (t \geq 1), \end{cases} \tag{2}$$

   where $n$ is the period parameter, and $\alpha\,(= 2/(1+n))$ is the weight. In our experiments, we use the same crossover rule as the above SMA for trades.

3. Bollinger Band (BB)
   The BB is an indicator based on the standard deviation of the change of prices, as follows:

$$\mathrm{BB}_n(t) = \mathrm{SMA}_n(t) \pm \beta\sigma, \tag{3}$$

$$\sigma = \sqrt{\frac{1}{n-1} \sum_{i=0}^{n-1} (c_{t-i} - \mathrm{SMA}_n(t-i))^2}, \tag{4}$$

   where $\beta$ is the factor of the standard deviation. The parameters of the BB are the period $n$ and the factor $\beta$.
   In our experiments, we apply the following rules: For a long trade, enter when the closing price crosses the upper line of the BB and exit when the closing price crosses $\mathrm{SMA}_n(t)$. For a short trade, enter when the closing price crosses the lower line of the BB and the exit rule is the same as the long one.

4. Price Channel Breakout (PCB)
   The PCB is an indicator based on the trading range of the last $n$ days, as follows:

$$U_n(t) = \max\{h_{t-i} | 1 \leq i \leq n\}, \tag{5}$$

$$L_n(t) = \min\{l_{t-i} | 1 \leq i \leq n\} \tag{6}$$

where $U_n(t)$ is the upper bound of the PCB, $L_n(t)$ is the lower bound, $h_t$ is the highest price and $l_t$ is the lowest one at Day $t$. The PCB has one parameter, namely the period $n$.

In our experiments, we apply the following rules: For a long trade, enter when $c_t > U_n(t)$ and exit when $l_t < \frac{1}{2}(U_n(t) + L_n(t))$. For a short trade, enter when $c_t < L_n(t)$ and exit when $h_t > \frac{1}{2}(U_n(t) + L_n(t))$.

## 5    Experiments

### 5.1    Setups

We apply our method on ten years of price data from the first trading day of 1999 to the last trading day of 2008. We selected twenty companies at random from the components of the Nikkei 225, which is a stock market index for the first section of the Tokyo Stock Exchange, and these issues are used for our experiments.

A sliding-window technique is applied for our experiments, as shown in Figure 7. A window consists of training and test. The former is a genetic search for four years. This search is applied to find effective indicators and their parameters for stock trading. The latter is a trading test for six months just after the training. This test is applied for the evaluation of the indicators and their parameters which are found by the genetic search. We have twelve windows whose start days slide six months in the ten years.

The initial principal is 5,000,000 JPY and the trading unit is minimal, *i.e.*, a round lot of each issue. A commission of one trade is assumed at 1,000 JPY. The rate to take profit in exit flags is 20% and the rate to stop loss is 10%. The limit of holding periods in exit flags is 50 days.

In Table 1, we show the technical indicators employed for our experiments. Our experiments use daily price data and every technical indicators are calculated from daily prices. The period for each indicator in the indirect coding takes

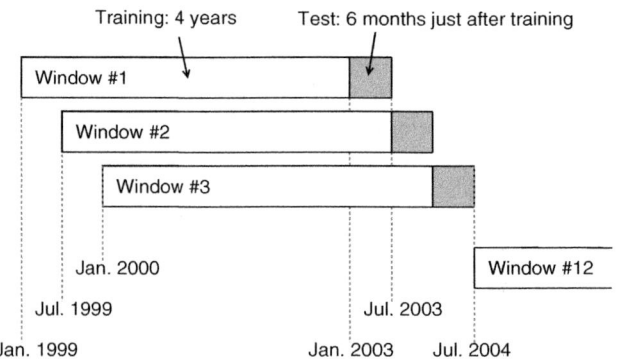

**Fig. 7.** Concept of sliding window

**Table 1.** Technical Indicators

| Indicators | Parameters |
|---|---|
| Simple Moving Average (SMA) | shorter period, longer period |
| Exponential Moving Average (EMA) | shorter period, longer period |
| Bollinger Band (BB) | period, factor |
| Price Channel Breakout (PCB) | period |

a value from a set of $\{5, 10, 15, 20, 25, 30, 50, 75, 100, 200\}$, and the factor for the Bollinger band also takes a value from a set of $\{1.0, 1.5, 2.0, 2.5, 3.0\}$. Since SMA has two parameters, 45 IDs are defined as *5-and-10 days SMA, 5-and-15 days SMA, ..., 100-and-200 days SMA*. In same ways, 45 EMAs, 50 BBs and 10 PCBs are also defined. Thus, the total number of IDs, which are combined indicators with parameters, is 150. In our genotype representation, each indicator can be applied for *long* and *short* positions. Therefore, the total size of IDs is 300 finally in the indirect coding.

On the other hand, in the direct coding, we use 8-bits binary coding method. The period for each indicator can take a value from 1 to 256 days and the range of the factor for BB is $[1.0, 3.0]$.

The setups of our genetic searches are the following: the population size is 50 and the searching generation is 5,000. The minimal generation gap model [9] is used for selection strategy. The uniform crossover is applied for recombination of individuals with 100% crossover probability. The random-replace mutation is used for the allele-based representation and the bit-flip mutation is done for the locus-based one. The mutation probability is $1/L$, where $L$ is the length of chromosomes, in both mutation operators. The fitness of each individual is the total interest obtained in the training period.

In our experiments, we compare three methods: the allele-based indirect representation, the locus-based indirect representation, and the locus-based direct representation. The first two methods are our proposed ones, whereas the third one corresponds to the conventional direct coding ([5] and [6]) described in Section 3. The objective of our experiments is to clarify some advantages of our proposed methods in comparing with the conventional one, the locus-based direct representation.

**Table 2.** Empirical results

| parameter coding | indirect | indirect | direct |
|---|---|---|---|
| indicator coding | allele-based | locus-based | locus-based |
| Num. of trades | 227 | 171 | 186 |
| Total profit | 2,427 | 1,710 | 1,628 |
| Worst draw down | 374 | 390 | 285 |
| Avg. CPU time | 1min. 14sec. | 2min. 08sec. | 16min. 12sec. |

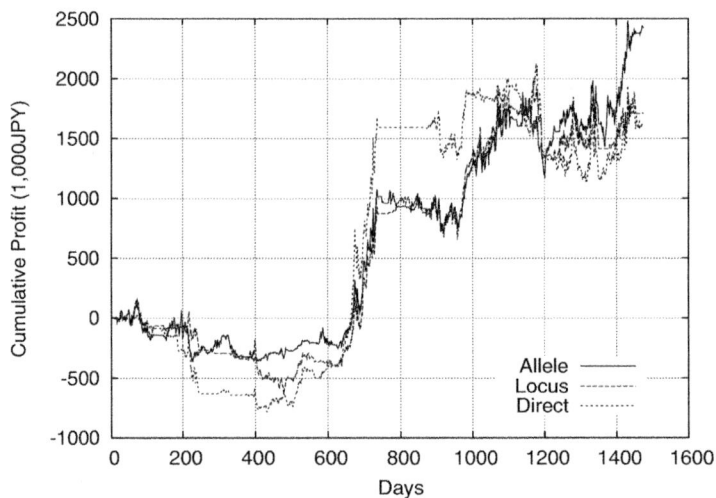

**Fig. 8.** Cumulative profit through our experiments

## 5.2   Results

We show the obtained results from our experiments through the twelve windows in Table 2. The number of trades is the total number of all the trades in the twelve windows. The profit and the draw down are represented in 1,000 JPY. The worst draw down is the worst loss from a series of loss trades. The average CPU time is the average of computational time for one window of our experiments in which we used Intel Xeon 3.0GHz.

In this table, the profit from the allele-based indirect representation is largest and the computational cost of this method is lowest in the three methods. The difference of the CPU time between the indirect and direct coding is extremely large.

We show the progress of cumulative profits of the three methods in Figure 8. Although all the methods suffered loss in earlier days, the loss was recovered later. We show the detailed results in every windows for the three methods in Tables 3, 4, and 5. Since the profit in each window changes dramatically, their standard deviations are very large.

In Tables 6, 7, and 8, we summarize the comparison of results between training and test. Since the training was applied for four years and the test was done for six months, the both results are converted into the value of one year and averaged over the twelve windows. The ratio in these tables is the ratio of the test to the training. In the three tables, the profits in the test are much reduced from that in the training.

**Table 3.** Results of allele-based indirect representation for twelve windows

| Window | Testing period | Num. of trades | % of wins | Profits | Worst draw down |
|---|---|---|---|---|---|
| 1 | 2003-1st | 32 | 34.4 | -140 | 87 |
| 2 | 2003-2nd | 34 | 26.5 | -144 | 252 |
| 3 | 2004-1st | 20 | 25.0 | -39 | 126 |
| 4 | 2004-2nd | 12 | 50.0 | 51 | 27 |
| 5 | 2005-1st | 19 | 52.6 | 50 | 86 |
| 6 | 2005-2nd | 11 | 90.9 | 1,238 | 27 |
| 7 | 2006-1st | 11 | 27.3 | -101 | 114 |
| 8 | 2006-2nd | 24 | 41.7 | 288 | 374 |
| 9 | 2007-1st | 14 | 50.0 | 477 | 118 |
| 10 | 2007-2nd | 30 | 30.0 | -125 | 291 |
| 11 | 2008-1st | 8 | 25.0 | 67 | 150 |
| 12 | 2008-2nd | 12 | 41.7 | 805 | 159 |
| Average | | 18.9 | 41.3 | 202.3 | 150.9 |
| Standard deviation | | 8.7 | 17.9 | 416.1 | 100.7 |

**Table 4.** Results of locus-based indirect representation for twelve windows

| Window | Testing period | Num. of trades | % of wins | Profits | Worst draw down |
|---|---|---|---|---|---|
| 1 | 2003-1st | 22 | 40.9 | -62 | 80 |
| 2 | 2003-2nd | 19 | 10.5 | -205 | 199 |
| 3 | 2004-1st | 4 | 0.0 | -71 | 71 |
| 4 | 2004-2nd | 31 | 35.5 | -159 | 172 |
| 5 | 2005-1st | 9 | 33.3 | 107 | 53 |
| 6 | 2005-2nd | 10 | 100.0 | 1,265 | 0 |
| 7 | 2006-1st | 6 | 50.0 | 8 | 33 |
| 8 | 2006-2nd | 22 | 45.5 | 399 | 340 |
| 9 | 2007-1st | 12 | 50.0 | 489 | 97 |
| 10 | 2007-2nd | 16 | 25.0 | -433 | 390 |
| 11 | 2008-1st | 15 | 26.7 | 77 | 135 |
| 12 | 2008-2nd | 5 | 60.0 | 295 | 141 |
| Average | | 14.3 | 39.8 | 142.5 | 142.6 |
| Standard deviation | | 7.8 | 24.4 | 420.6 | 113.9 |

## 6   Discussion

In our experiments, the allele-based indirect representation obtained the largest profit and its computational cost was lowest in the three methods, as shown in Table 2. From this result, it turned out that the allele-based indirect representation was very effective for automatic stock trading using genetic algorithms.

The difference of the computational costs was extremely large because the indirect coding needs only restricted sets of values for parameters whereas the direct coding needs very large sets of parameter values. The low computational cost is an advantage of our method.

In Figure 8 and Tables 3, 4, 5, the profit in each window of our experiments changed dramatically. In particular, some earlier sets suffered losses. However, in the allele-based indirect representation, the early loss is less than the others

**Table 5.** Results of locus-based direct representation for twelve windows

| Window | Testing period | Num. of trades | % of wins | Profits | Worst draw down |
|--------|----------------|----------------|-----------|---------|-----------------|
| 1 | 2003-1st | 10 | 40.0 | -85 | 49 |
| 2 | 2003-2nd | 37 | 13.5 | -545 | 235 |
| 3 | 2004-1st | 1 | 0.0 | -11 | 11 |
| 4 | 2004-2nd | 12 | 8.3 | -95 | 157 |
| 5 | 2005-1st | 23 | 47.8 | 332 | 72 |
| 6 | 2005-2nd | 13 | 92.3 | 1,999 | 2 |
| 7 | 2006-1st | 0 | 0.0 | 0 | 0 |
| 8 | 2006-2nd | 19 | 57.9 | 279 | 268 |
| 9 | 2007-1st | 12 | 50.0 | 68 | 189 |
| 10 | 2007-2nd | 21 | 23.8 | -620 | 285 |
| 11 | 2008-1st | 22 | 31.8.0 | -155 | 238 |
| 12 | 2008-2nd | 16 | 56.3 | 461 | 282 |
| Average | | 15.5 | 35.1 | 135.7 | 149.0 |
| Standard deviation | | 9.6 | 26.5 | 639.2 | 110.3 |

**Table 6.** Comparison between training and test: allele-based indirect representation

| | Training | Test | Ratio |
|--|----------|------|-------|
| Avg. num. of trades | 48.5 | 37.8 | 78% |
| Avg. profit | 1038 | 405 | 39% |
| Avg. worst draw down | 312 | 151 | 48% |

**Table 7.** Comparison between training and test: locus-based indirect representation

| | Training | Test | Ratio |
|--|----------|------|-------|
| Avg. num. of trades | 43.9 | 28.6 | 65% |
| Avg. profit | 1006 | 285 | 28% |
| Avg. worst draw down | 311 | 143 | 46% |

**Table 8.** Comparison between training and test: locus-based direct representation

| | Training | Test | Ratio |
|--|----------|------|-------|
| Avg. num. of trades | 40.2 | 31.0 | 77% |
| Avg. profit | 1065 | 271 | 25% |
| Avg. worst draw down | 243 | 149 | 61% |

**Table 9.** Detailed results in the latter half of 2008 by the allele-based indirect representation

|  | Long trades | Short trades |
|---|---|---|
| Obtained IDs | SMA (50/75 days) | SMA (25/30 days) BB (50days 1.0) |
| Num. of trades | 2 | 10 |
| Win/Loss | 0/2 | 5/5 |
| Avg. profit | -71.0 | 94.7 |
| Avg. profit for win | 0.0 | 210.4 |
| Avg. profit for loss | -71.0 | -21.0 |

in the days 200 to 600 in Figure 8. This indicates that our method is robust in variable markets for stock trading.

The allele-based indirect representation obtained larger profit than the others in the twelfth window, which is the latter half of 2008, as shown in Tables 3, 4, and 5. The annual interest rate of this method is 32.2% ($= 2 \times 805 \div 5000$). This is remarkable in the global financial crisis which has occurred in the latter half of 2008.

In Table 9, we show the detailed trading results and the obtained IDs, which are the combined indicator and its parameters, from the allele-based indirect representation in the twelfth window. For short trades, our system employs the conjunction of the SMA and the BB. The obtained profit was derived from the short trades. The average profit of the winning short trades was much larger than that of the loss trades. This is the reason that the large profit was obtained even though the winning percentage was 50% in short trades.

In Tables 6, 7, and 8, the average profits for the test were much less than that for the training. This implies that the obtained indicators and the parameters were overfitted to the training sets and it was hard to fit the test sets.

We also show the average of the worst draw downs in the twelve windows in Tables 6, 7, and 8. The average of the worst draw downs on the test sets is less than that on the training sets. Since it is important to reduce the worst draw downs for constructing steady trading systems, this characteristic of our system is remarkable.

## 7   Conclusions

We proposed the allele-based indirect representation for acquiring stock trading strategy using genetic algorithms and we compared three types of genotype representations empirically. In our experiments, the allele-based indirect representation outperformed the other two methods, which includes the conventional locus-based direct representation. In particular, the indirect coding was superior to the direct coding in computational costs, and our method obtained large profit in the global financial crisis in 2008.

Future problems are the following: To keep off the overfitting in the training and to reduce the fluctuations of profits through windows of experiments. Also, it is important to add other technical indicators since we used only four types of indicators in our experiments.

# References

1. Brabazon, A., O'Neill, M.: An Introduction to Evolutionary Computation in Finance. IEEE Computational Intelligence Magazine 3, 42–55 (2008)
2. Goldberg, D.E.: Genetic Algorithms in Search, Optimization and Machine Learning. Addison-Wesley, Reading (1989)
3. Dempster, M.A.H., Jones, C.M.: A Real-time Adaptive Trading System Using Genetic Programming. Quantitative Finance 1, 397–413 (2001)
4. Hryshko, A., Downs, T.: An Implementation of Genetic Algorithms as a Basis for a Trading System on the Foreign Exchange Market. In: Proc. Congress of Evolutionary Computation, pp. 1695–1701 (2003)
5. de la Fuente, D., Garrido, A., Laviada, J., Gomez, A.: Genetic Algorithms to Optimise the Time to Make Stock Market Investment. In: Proc. of Genetic and Evolutionary Computation Conference, pp. 1857–1858 (2006)
6. Hirabayashi, A., Aranha, C., Iba, H.: Optimization of the Trading Rule in Foreign Exchange using Genetic Algorithms. In: Proc. of the 2009 IASTED Int'l Conf. on Advances in Computer Science and Engineering (2009)
7. Eshelman, L.J., Schaffer, J.D.: Real-coded Genetic Algorithms and Interval-Schemata. Foundations of Genetic Algorithms 2, 187–202 (1993)
8. Tsutsui, S., Yamamura, M., Higuchi, T.: Multi-parent Recombination with Simplex Crossover in Real Coded Genetic Algorithms. In: Proc. of Genetic and Evolutionary Computation Conference, pp. 657–664 (1999)
9. Satoh, H., Yamamura, M., Kobayashi, S.: Minimal Generation Gap Model for GAs Considering Both Exploration and Exploitation. In: Proc. of 4th Int'l Conf. on Soft Computing, pp. 494–497 (1996)

# Automatic Adaptive Modeling of Fuzzy Systems Using Particle Swarm Optimization

Sergio Oliveira Costa Jr.[1], Nadia Nedjah[1], and Luiza de Macedo Mourelle[2]

[1] Department of Electronics Engineering and Telecommunications
[2] Department of Systems Engineering and Computation
Faculty of Engineering, State University of Rio de Janeiro
{serol,nadia,ldmm}@eng.uerj.br

**Abstract.** Fuzzy systems are currently used in many kinds of applications, such as control, for their effectiveness and efficiency. However, these characteristics depend primarily on the model yield by human experts, which may or may not be optimized for the problem at hand. Particle swarm optimization (PSO) is a technique used in complex problems, including multi-objective problems. In this paper, we propose an algorithm that can generate fuzzy systems automatically for different kinds of problems by simply providing the objective function and the problem variables. This automatic generation is performed using particle swarm optimization. To be able to do so and in order to avoid dealing with inconsistent fuzzy systems, we used some known techniques, such as the WM method, to help in developing meaningful rules and clustering concepts to generate membership functions. Tests using three three-dimensional functions have been carried out and the obtained results are presented.

## 1 Introduction

Fuzzy systems [5] form an important tool to model complex problems based on imprecise informations and/or in situations where a precise result is not of interest and an approximation is sufficient [16]. The performance of a fuzzy system depends on the expert's interpretation, which leads to in the generation of the rule base and membership functions of the system. To minimize this dependency, some methods are being used in the attempt to automatically generate the components required in a fuzzy system. For the membership functions, clustering-based algorithms, such as Fuzzy C-means and its generalizations as *Pre-shaped C-means* [2], are usually used. Other approaches also exist [10]. The major difficulty in the development of fuzzy systems consists of the definition of membership functions and rules that provide the desired behavior of these systems.

Swarm Intelligence is an area of artificial intelligence based on collective and decentralized behavior of individuals that interact with each other, as well as with the environment [1]. PSO is a stochastic evolutionary algorithm, based on swarm intelligence, that searches for the solution of optimization problems in a specific search space and is able to predict the social behavior of individuals according to defined objectives [6].

M.L. Gavrilova et al. (Eds.): Trans. on Comput. Sci. VIII, LNCS 6260, pp. 71–84, 2010.

Methods based on examples, such as the Wang-Mendel or WM method [15], are usually used for automatic rule generation. Also, there are many research works that exploit evolutionary algorithms (EA), both to optimize the rule base and the membership functions. In [3], genetic algorithms (GA) are used to generate the rule base, with candidate rules pre-selection. In [9], the authors use EA to generate fuzzy systems that are more compact and more interpretable by humans. In [13], the authors use clustering techniques and GA to define good sets of rules for classification problems. In [4], the authors use evolutionary technique and GA to generate fuzzy systems from some given knowledge bases.

In this paper, we developed an algorithm based on PSO to generate fuzzy systems for any kind of problem, provided an objective function that may be continuous or discrete. Using simple informations, such as variable names, the corresponding domains and the objective function, this algorithm can yield an appropriate fuzzy system. Some tests were performed with a known control surface to validate the effectiveness of the tool.

The rest of this paper is organized in five sections. Firstly, in Section 2, we explain briefly the principle behind PSO. Then, in Section 3, we describe the WM method of rules generation. After that, in Section 4, we give details about the proposed method for the automatic modeling of fuzzy systems using PSO. For this purpose, we first define the structure of a particle and the coordinates used to position it within the search space as well as the fitness function of the represented system. Then, in Section 5, we present the obtained results to model a commonly used control surface. Last but not least, in Section 6, we conclude the reported work and give some directions for future research.

## 2    Particle Swarm Optimization

During a particle swarm optimization process, each particle is mapped into a position in the search space, which $n$-dimensional. The particle position is updated in each iteration. For the position update particle $i$, the velocity related to each of the search space directions is used. The velocity is the element that promotes the movement of the particles and is calculated as in (1) [11,8,6].

$$v_i(t+1) = wv_i(t) + c_1 r_1(\hat{x}_i(t) - x_i(t)) + c_2 r_2(\bar{x}(t) - x_i(t)), \tag{1}$$

where $w$ is called inertia coefficient, $r_1$ and $r_2$ are random numbers chosen in the interval [0,1], $c_1$ and $c_2$ are positives constants called as social and cognitive coefficients, $\hat{x}_i(t)$ identifies the best position achieved by the particle in the past and $\bar{x}(t)$ is the best position, among all the particles, achieved in the past. The position of the particle is updated as described in (2).

$$x_i(t+1) = x_i(t) + v_i(t+1). \tag{2}$$

The velocity guides the optimization process [6], reflecting both the particle experience and the information exchange between particles. The experimental knowledge of a particle refers to the cognitive component, which is proportional to the distance between the particle and its best position, found so far. The information exchange between particles refers to the social component of the velocity equation (1).

To avoid that a particle leaves the search space, it is necessary to use a parameter that bounds the velocity [6]. This is known as the maximum velocity $v_{max}$, it allows a higher granularity on the search control. Therefore, before executing the update defined in (2), the velocity is analyzed with respect to the criterion defined in (3).

$$
v_i(t+1) =
\begin{cases}
v_i(t+1) \text{ if } v_i(t+1) < v_{max} \\
\\
v_{max} \quad \text{ if } v_i(t+1) \geq v_{max}
\end{cases}
\tag{3}
$$

In (1), one can observe three terms that interfere in the velocity computation [6], which are:

- The *previous velocity*, $wv_i(t)$, is used to prevent particle $i$ to suffer a drastic change in direction. This component also is called of inertia component.
- The *cognitive component*, $c_1 r_1(\hat{x}_i(t) - x_i(t))$, quantifies the performance of particle $i$ with respect to previous performances. This component was defined by Kennedy and Eberhart as the "nostalgia" of the particle [8].
- The *social component*, $c_2 r_2(\bar{x}(t) - x_i(t))$, quantifies the performance of particle $i$ with respect to the performance of the set of included particles. The effect of this term is to attract the particle to the best position found by the particles set.

The value assigned to each parameter of PSO algorithm is essential in the search process evolution. Below are related some set of values considered *good*.

- The *inertia coefficient*, $w$, controls the relation between exploration and exploitation [14]. The values near 1 are considered good, but values bigger than 1 are not so good so are very small values [6]. Values bigger than 1 tend to leave the particles with a very high acceleration, promoting a high divergence, while very small values can make the search too slow.
- The *cognitive coefficient*, $c_1$, and the *social coefficient*, $c_2$, yield a better performance when these are balanced, i.e., $c_1 \cong c_2$ [6].
- The factors $r_1$ and $r_2$ define the stochastic nature of the cognitive and social contributions. Random values are selected in the range [0,1] [6] to each factors.
- The *maximum velocity* is defined for each of the dimensions of search space and can be formulated as a domain percentage [6], $v_{max} = \delta(x_{max} - x_{min})$, where $x_{max}$ and $x_{min}$ are the maximum and minimum domain values respectively and $\delta$ is a value in the range [0, 1].
- The *number of particles* define the possibility to cover a range of the search space in every iteration of the algorithm. A high number of particles allows a better coverage of the search space but requires a considerable computational power. Empirical studies show that PSO achieves optimal solutions using ten to thirty particles [6].

Let $pBest$ be the best position found by a particle and $gBest$ be the best position among those found by all the particles. Algorithm 1 is describes the PSO optimization process. A given maximal iteration number and the predefined fitness values can be used as a stop criterion.

---

**Algorithm 1.** Particle swarm-based optimization algorithm (PSO)

---

1:  **for** $i := 1$ **until** *total_particulas* **do**
2:      Initialize particle $i$ information;
3:      Initialize random position of particle $i$;
4:      Initialize random velocity of particle $i$;
5:  **end for**
6:  **repeat**
7:      **for** $i := 1$ **until** *total_particulas* **do**
8:          Calculate fitness of particle $i$;
9:          **if** (fitness better than $pBest_i$) **then**
10:             Update $pBest_i$ with the new position;
11:         **end if**
12:         **if** (fitness better than gBest) **then**
13:             Update gBest with the new position;
14:         **end if**
15:         Update velocity of particle $i$;
16:         Update position of particle $i$;
17:     **end for**
18: **until** ($stopcriterion = true$)

---

## 3   Rule Generation Methods

The rule generation method referred to as Wang-Mendel (WM) [15] uses an input-output data set for the problem at hand, to generate a rule set of fuzzy systems. The input-output data set is usually provided as $(x^p; y^p)$, $p = 1, 2, \ldots, N$, wherein $x^p \in R^n$ and $y^p \in R$. This method extracts the rules that best describe how the output variable $y \in R$ is influenced by the $n$ input variables $x = (x_1, ..., x_n) \in R^n$, based on the provided examples.

For instance, assuming two data sets to a system with two input variables $x_1$ and $x_2$ and an output $y$, that are $(x_1^1, x_2^1, y^1)$ and $(x_1^2, x_2^2, y^2)$, and the membership functions showed in the graphics of the Fig. 1. To obtain the rules represented by these two sets, first we must get the degree of confidence using the membership functions, for each data set. In this case, we have:

- $x_1^1$: degree 0.67 in $A1$ and 0.11 in $A2$;
- $x_2^1$: degree 0.16 in $B1$ and 0.80 in $B2$;
- $y^1$: degree 0.66 in $C1$;
- $x_1^2$: degree 0.25 in $A2$ and 0.68 in $A3$;
- $x_2^2$: degree 0.10 in $B3$ and 0.58 in $B4$;
- $y^2$: degree 0.39 in $C2$ and 0.51 in $C3$.

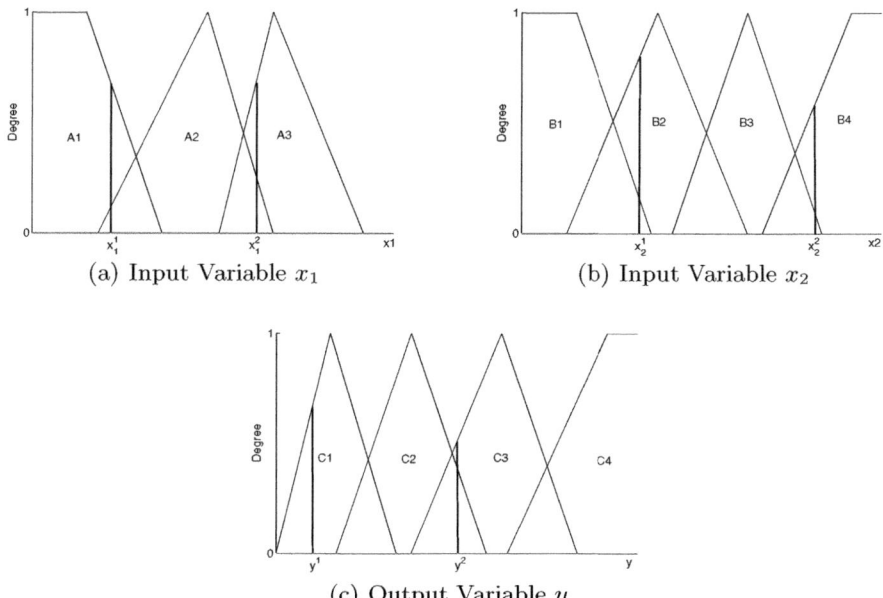

(a) Input Variable $x_1$    (b) Input Variable $x_2$

(c) Output Variable $y$

**Fig. 1.** Example of input-output data set for rules generation

In order to generate the rules, we always keep the membership functions in which the variable has the highest degree, and so we discard the functions that have lower degree. So, for the data sets defined in Fig. 1, we have (4). So the first rule would be "If $x_1$ is $A1$ and $x_2$ is $B2$ then $y$ is $C1$" and the second, "If $x_1$ is $A3$ and $x_2$ is $B4$ then $y$ is $C3$".

$$(x_1^1, x_2^1, y^1) = [x_1^1(0.67 \text{ in } A1), x_2^1(0.80 \text{ in } B2), y^1(0.66 \text{ in } C1)]$$
$$x_1^2, x_2^2, y^2) = [x_1^2(0.68 \text{ in } A3), x_2^2(0.58 \text{ in } B4), y^2(0.51 \text{ in } C3)] \qquad (4)$$

Note that each data set generates one single rule. Considering a real system, it is very possible that these rules can be conflicting rules. To overcome this problem, one can associate degrees of confidence to each generated rule, using the degree of relevance of each rule term. Equation 5 shows how this degree can be computed:

$$C(Rule) = \mu(x_1) \times \mu(x_2) \times \mu(y), \qquad (5)$$

wherein $C$ is degree of $Rule$ and $\mu(x1)$, $\mu(x2)$ and $\mu(y)$ are the degree of relevance of each rule term. In the case of the first rule, the associated confidence degree would be $C(Rule_1) = 0.67 \times 0.80 \times 0.66 = 0.35376$.

There are also methods based on genetic algorithms. In [3], the authors use the WM method to generate the initial rule set and then apply their own genetic algorithm on some classification rate of the rules. This method is only used for classification problems.

## 4   Proposed Automatic Generation

In this work, PSO is used to evolve the fuzzy systems parameters of the *Mamdani* type [5], both for rules and membership functions. The search algorithm is based on these two elements and always tries to improve the solution at hand. However, the functions are not modified in every iteration, unlike the rules, whose modification obeys to pre-determined update rate, that is defined at the beginning of the evolutionary process. The purpose is to maintain the functions stable for some time, giving more time for the algorithm to search for more appropriate rules for those functions. At the end of each execution, when the algorithm reaches the stop criterion, it returns the best solution found.

There are four important aspect that define the performance of the PSO search. These are the particle representation, the position coordinates of a given particle in the search space, the fitness function that allows us to determine how appropriate is the fuzzy system associated with a given particle and how to update the system represented by the particle at hand.

### 4.1   Representation

Each particle is associated with a fuzzy system and a position in the search space, that is represented by an $n$-position vector, where $n$ depends on the number of used variables. In this work, the fuzzy system is defined by an hierarchical structure as described in Fig. 2.

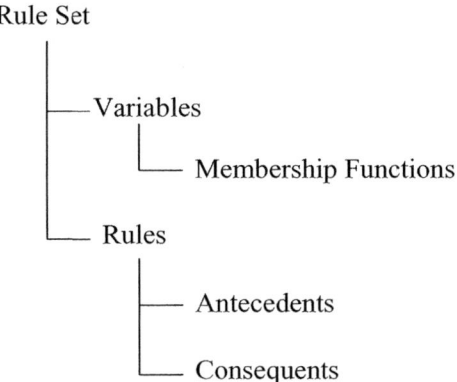

**Fig. 2.** Fuzzy representation structure

### 4.2   Particle Position and Movement

The position vector has one entry for the rules number, another one for the completeness factor, and $m$ positions for the number of the functions, where $m$ is the number of the system variables. Thus, the position dimension is dependent on the number of variables of the problem. The completeness factor is a criterion that measures the *discontinuity* between functions in the variables domain (see Section 4.3). The mutation operator determines how the update of each one these items is performed. This update promotes the movement of particles on the search space. In this work, we used three kinds of mutations:

– If the velocity relating to the number of rules is positive, then we increase the rules number. Otherwise, we decrease it.
– Changing the number of functions for each variable of the fuzzy system follows the same criterion, given above.
– The change of the completeness is performed increasing the width of a function. Thus, the tendency is to reduce the empty space in the domain, if any. Similarly, reducing the width of the function, we alter the distinctness between the available ones. The more positive the velocity is, the bigger the increase in domain of one of the functions. If the more negative the velocity is, the smaller the decrease in the domain.

## 4.3   Fitness Function

Inspired by the work reported in [9], the fitness of each particle is defined as in (6):

$$\begin{aligned}
F = \ &-100 \times \omega_1 \times \log(MSE) \\
&+50 \times \omega_2(1 - C_r) \\
&+50 \times \omega_3(1 - C_f) \\
&+50 \times \omega_4(1 - P_D) \\
&+50 \times \omega_5(1 - P_C),
\end{aligned} \tag{6}$$

wherein $MSE$ is the mean-square error of the difference between the returned value by the objective function ($y_h$) and the returned value by the fuzzy model ($y_h^F$), as (7), wherein $N_D$ is the number of data.

$$MSE = \frac{1}{N_D} \sum_{h=1}^{N} (y_h - y_h^F)^2, \tag{7}$$

The $C_r$ term represents the relation between the amount of rules presents on the model and the total of possible rules, and $C_f$ represents the relation between the amount of functions presents on the system and the total of possible functions, as in (8):

$$C_r = \frac{N_R}{N_R^{max}}$$
$$C_f = \frac{N_F}{N_F^{max}} \tag{8}$$

Term $P_D$ is a criterion that measures the *distinctness* between the membership functions of the variables, defined in (9):

$$P_D = \frac{1}{N_V} \sum_{i=1}^{N_V} \left( \frac{1}{N_S^i} \sum_{h=1}^{N_S^i} \frac{\lambda_{ih}}{|\chi_i|} \right), \tag{9}$$

wherein $N_V$ is the number of variables, $N_S^i$ is the total possible interval of overlap between functions of the i-th variable, $\lambda_{ih}$ is the width of the h-th overlap and $\chi_i$ is the width of the variable domain. In order to the determine $\lambda_{ih}$, it is necessary to define the level $\xi_D$, drawing a horizontal line, crossing all the functions, as showed in the Fig. 3(a). Term $P_C$ is a criterion that represents the *completeness* of the membership functions, in relation to the domain, and is defined as in (10):

$$P_C = \frac{1}{N_V} \sum_{i=1}^{N_V} \left( \frac{1}{N_D^i} \sum_{h=1}^{N_D^i} \frac{\gamma_{ih}}{|\chi_i|} \right), \tag{10}$$

wherein $N_D^i$ is the total possible number of discontinuity between functions of the $i$-th variable, $\gamma_{ih}$ is the width of $h$-th discontinuity and $\chi_i$ is the width of the variable domain.

In order to determine $\gamma_{ih}$, is necessary to define the level $\xi_C$, in a similar way to $\xi_D$, as shown Fig. 3(b).

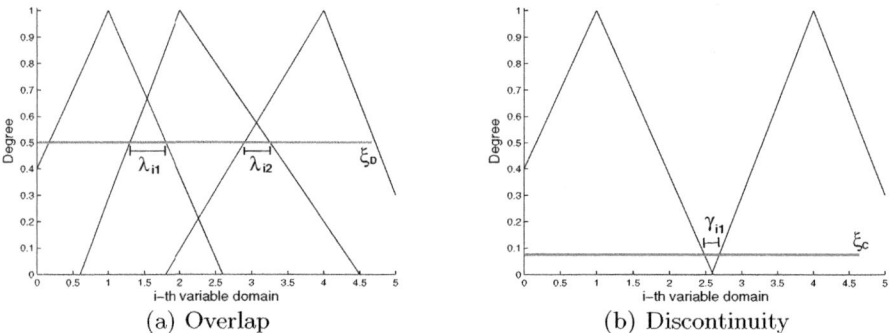

(a) Overlap                              (b) Discontinuity

**Fig. 3.** Overlap and discontinuity illustrations

The coefficients $\omega_1$, $\omega_2$, $\omega_3$, $\omega_4$ and $\omega_5$, control the contribution of each term of (6) in the evaluation of the fuzzy system associated with the particle.

This evaluation function covers the many required criteria of a fuzzy system. These are the *precision*, by the error quantification; the *compactness*, by the relation between the number of the rules and functions of the model and the total possible number; and the *interpretability*, by measuring of the distinctness and completeness, providing a complete model evaluation.

## 5   Results and Tests

The WM method [15], introduced in Section 3, was used to initialize rules of the fuzzy systems of each particle. Besides, the concept of clustering was used in the membership functions generation, to decrease the possibility of yielding functions that are incompatible with the fuzzy system. In order to evaluate the implementation, reported throughout this paper, we performed some tests to generate the control surface of a water vehicle [7] shown in Fig. 4(a). Equation (11) shows the function used to generate this curve.

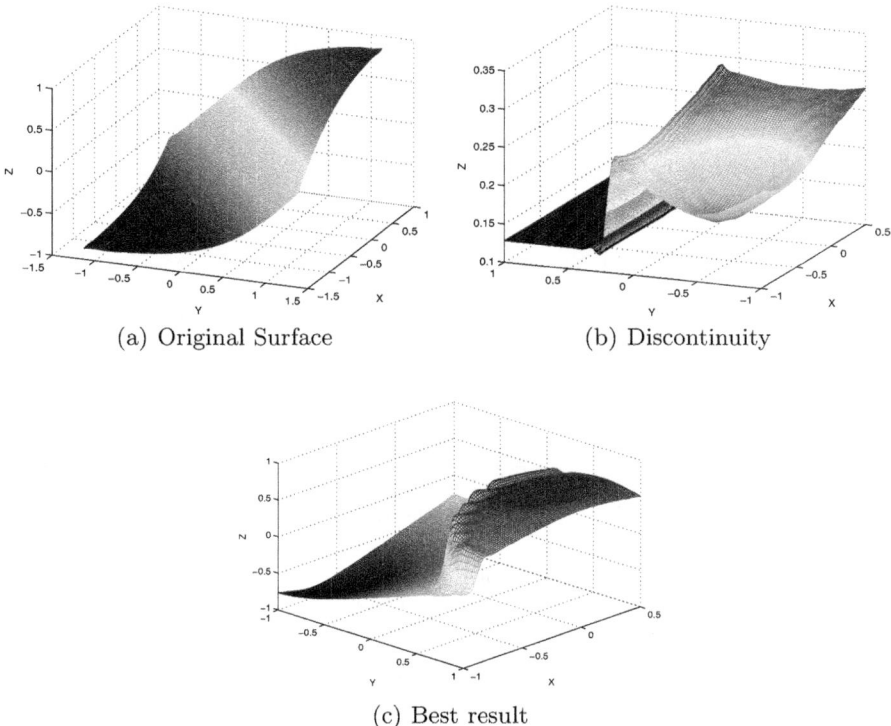

(a) Original Surface          (b) Discontinuity

(c) Best result

**Fig. 4.** Original surface and an initial result

**Table 1.** Algorithm parameters

| Parameter | Value |
|---|---|
| Inertia coefficient | 0, 0.5, 1 |
| Social and cognitive coefficient | 1, 1.1, 1.5, 1.9, 2, 3, 4 |
| Number of particles | 10, 20, 40, 100 |
| Total of iterations | 1000, 10000, 50000 |
| Minimum number of rules | 1, 2, 4 |
| Maximum number of rules | 1, 4, 16, 50 |
| Minimum number of functions | 1, 2 |
| Maximum number of functions | 2, 4, 7 |
| Kind of Function | different kind of functions<br>Gaussian functions only<br>Triangular functions only |
| Initialization | With WM e Without WM |
| Total Rules WM | 0, 2 e 4 |
| $\omega_1 - \omega_5$ | 0, 0.5, 1 |

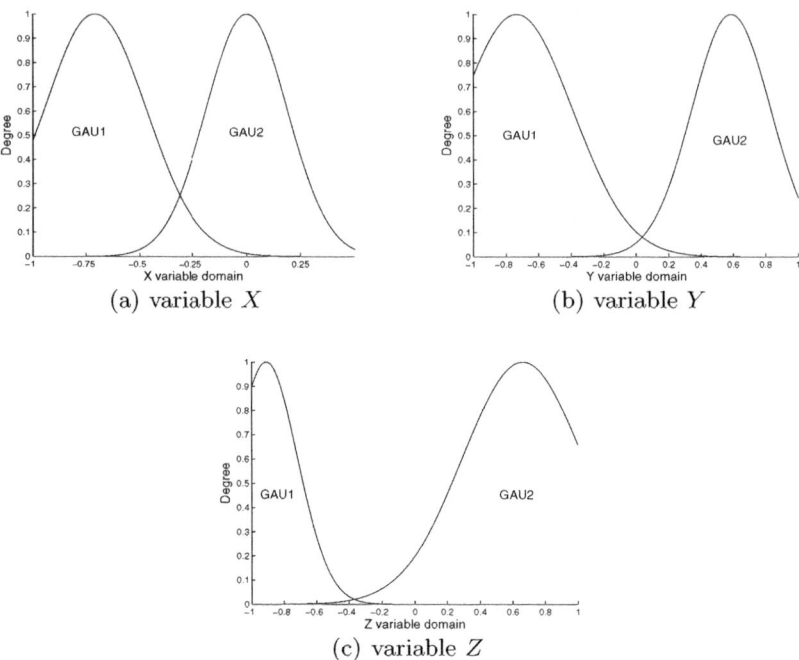

(a) variable $X$

(b) variable $Y$

(c) variable $Z$

**Fig. 5.** Functions of the variable $X$, $Y$ and $Z$

$$z = \frac{2}{1 + e^{-2x - 2y}} - 1 \qquad (11)$$

Initially, we used variable parameters: number of rules may vary from 1 to 50 and membership functions from 1 to 7 for each variable. However, the results were not satisfactory. The surfaces generated by the evolved fuzzy systems were too far from the original one, as shown in Fig. 4(b).

Several tests were performed, and the PSO parameters were adjusted many times with values near to those referenced in related literature [6]. Defining a small interval for the rules set and setting the values of the number of functions

**Table 2.** Values of the algorithm parameters

| Parameter | Value |
|---|---|
| Social coefficient | 1.5 |
| Cognitive coefficient | 1.5 |
| Inertia coefficient | 1 |
| Number of particles | 20 |
| Completeness | 0.25 |
| Number of iteration | 5000 |
| Total of rules WM | 6 |
| Kind of function | Gaussian |
| $\omega_1 - \omega_5$ | 1 |

**Table 3.** Comparison of the performance of GA-based *vs.* PSO-based fuzzy system automatic generation for functions $seno(xy)$ and $e^{xsin(\pi y)}$

| Function | Algorithm | Fitness | Deviation of fitness | Error | Deviation of error |
|---|---|---|---|---|---|
| $seno(xy)$ | GA | 54 | 8.0 | 0.0288 | 0.0007 |
| | PSO | 275.5 | 3.8 | 0.1059 | 0.0537 |
| $e^{xsin(\pi y)}$ | GA | 197 | 35.0 | 0.006 | 0.0032 |
| | PSO | 280.5 | 16.8 | 0.176 | 0.1656 |

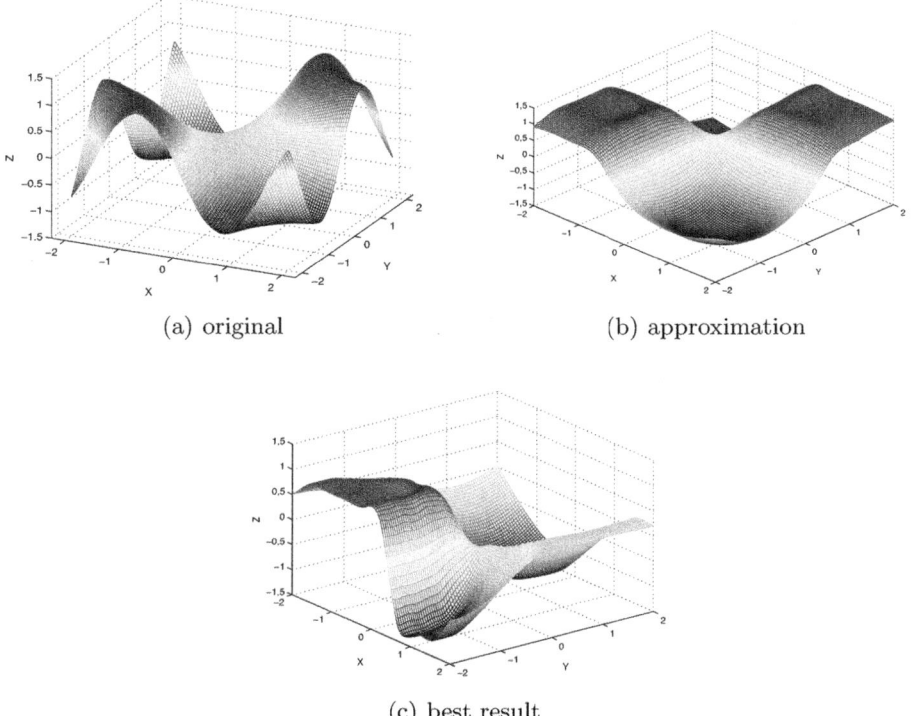

(a) original                (b) approximation

(c) best result

**Fig. 6.** Original function $z = seno(xy)$ and its approximations by the generated fuzzy systems

(number of rules varying from 1 to 4 and two membership functions for each variable), the algorithm evolved a fuzzy system that yielded curves that meet the original surface in several points. Table 1 shows some variations of parameters, used in the tests. The best result obtained so far is depicted in Fig. 4(c).

The set of rules evolved area as follows. The membership functions of the fuzzy systems that provided the best results are presented in Fig. 5(a), 5(b) e 5(c).

1. if $X$ is $GAU_1$ and $Y$ is $GAU_1$ then $Z$ is $GAU_1$;
2. if $X$ is $GAU_2$ and $Y$ is $GAU_2$ then $Z$ is $GAU_2$.

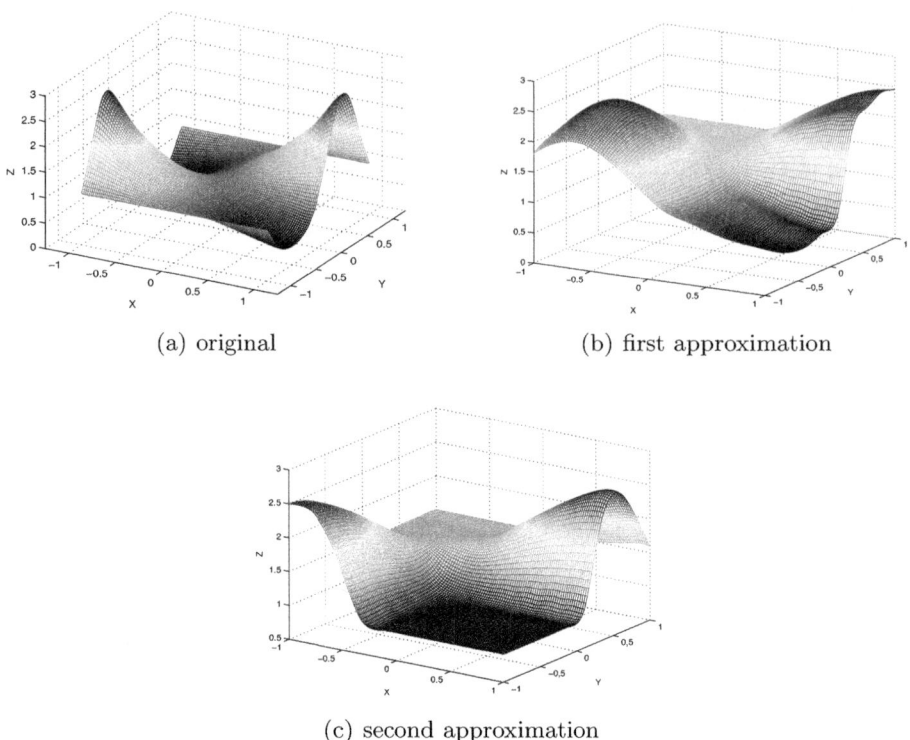

(a) original

(b) first approximation

(c) second approximation

**Fig. 7.** Original function $z = e^{x sin(\pi y)}$ and its approximations by the generated fuzzy systems

More experiments were performed. Table 2 show the used parameters by the PSO-based algorithm. The results obtained for the approximation of function $z = seno(xy)$ using the fuzzy system generated by the PSO-based algorithm are shown in Fig. 6, wherein Fig. 6(a) depicts the graphics of the original function. Fig. 6(b) and Fig. 6(c) show the best approximation results of function $z = seno(xy)$.

The results obtained for the approximation of function $z = e^{x sin(\pi y)}$ using the fuzzy system generated by the PSO-based algorithm are shown in Fig. 6, wherein Fig. 7(a) depicts the graphics of the original function. Fig. 7(b) and Fig. 7(c) show the best approximation results of function evolved for that function.

Table 3 shows a comparison between the results obtained by automatic generation of fuzzy systems using genetic algorithms [12] and using the proposed method using PSO. The error is computed using the Normalized Mean Square Error (NMSE).

# 6    Conclusion

In this paper, we illustrated the use of PSO to automatically generate the fuzzy rules, fuzzy variables together with the corresponding membership functions of fuzzy systems. We described the particle representation, its movement in the search space and we provided a fitness function that allows us to assess the appropriateness of the evolved fuzzy system. This experience showed that the performance of the evolutionary process is very much dependent on the choice of the parameters, such as the number of membership functions per variable as well as on the number of rules allowed in the system. More tests are being carried out in order to synthesize discrete functions into fuzzy systems.

## Acknowledgments

The authors are grateful to FAPERJ (*Fundação de Amparo à Pesquisa do Estado do Rio de janeiro*, http:// www.faperj.br), CNPq (*Conselho Nacional de Desenvolvimento Científico e Tecnológico*, http://www.cnpq.br) and CAPES (*Coordenação de Aperfeiçoamento de Pessoal de Nível Superior* http://www.capes.gov.br)for their continuous financial support.

## References

1. Beni, G., Wang, J.: Robots and Biological Systems: Towards a New Bionics? NATO ASI Series, Toscana, Italy (1989)
2. Chen, L., Chen, C.L.P.: Pre-shaped fuzzy c-means algorithm (pfcm) for transparent membership function generation. In: Proc. of IEEE International Conference on Systems, Man and Cybernetics, pp. 789–794 (October 2007)
3. Cintra, M.E., Camargo, H.A.: Fuzzy rules generation using genetic algorithms with self-adaptive selection. In: Proc. of IEEE International Conference on Information Reuse and Integration, pp. 261–266 (August 2007)
4. Cordón, O., Herrera, F.: A hybrid genetic algorithm-evolution strategy process for learning fuzzy logic controller knowledge bases. In: Genetic Algorithms and Soft Computing, pp. 251–278. Physica-Verlag, Heidelberg (1996)
5. Cox, E.: The Fuzzy Systems Handbook: A Practitioner's Guide to Building, Using, and Maintaining Fuzzy Systems. Academic Press Limited, Oval Road (1994)
6. Engelbrecht, A.P.: Fundamentals of Computational Swarm Intelligence. John Wiley & Sons Ltd., England (2005)
7. Guo, B., Liang, X., Wang, B., Wan, L.: Sigmoid surface control for mini underwater vehicles by improved particle swarm optimization. In: Proc. of International Conference on Robotics and Biomimetics (December 2007)
8. Kennedy, J., Eberhart, R.: Particle Swarm Optimization. In: Proc. of IEEE International Conference on Neural Networks, pp. 1942–1948. IEEE Computer Society Press, Los Alamitos (1995)
9. Kim, M.S., Kim, C.-H., Lee, J.j.: Evolving compact and interpretable takagi-sugeno fuzzy models with a new encoding scheme. IEEE Transactions on Systems, Man, and Cybernetics, Part B 36, 1006–1023 (2006)

10. Krone, A., Slawinski, T.: Data-based extraction of unidimensional fuzzy sets for fuzzy rule generation. In: Proc. of IEEE International Conference on Fuzzy Systems, vol. 02, pp. 1032–1037 (May 1998)
11. Nedjah, N., Mourelle, L.M.: Swarm Intelligent Systems. Springer, Heidelberg (2006)
12. Rivas, V.M., Merelo, J.J., Rojas, I., Romero, G., Castillo, P.A., Carpio, J.: Evolving two-dimensional fuzzy systems. Fuzzy Sets Systems 138(2), 381–398 (2003)
13. Setnes, M., Roubos, H.: GA-fuzzy modeling and classification: complexity and performance. IEEE Transactions on Fuzzy Systems 08, 509–522 (2000)
14. Shi, Y., Eberhart, R.C.: A Modified Particle Swarm Optimizer. In: Proc. of IEEE Congress on Evolutionary Computation, pp. 69–73. IEEE Computer Society Press, Los Alamitos (1998)
15. Wang, L.X.: The WM method completed: A flexible fuzzy system approach to data mining. IEEE Transactions on Fuzzy Systems 11, 768–782 (2003)
16. Zadeh, L.A.: Fuzzy sets. Information and Control 08, 338–353 (1965)

# Part II

**Computational Methods for Model Visualization
and Analysis**

# Computational Algorithm for Some Problems with Variable Geometrical Structure

N. Bessonov[1] and V. Volpert[2]

[1] Institute of Mechanical Engineering Problems, 199178 Saint Petersburg, Russia
[2] Institut Camille Jordan, UMR 5208 CNRS, University Lyon 1
69622 Villeurbanne, France

**Abstract.** The work is devoted to the computational algorithm for a problem of plant growth. The plant is represented as a system of connected intervals corresponding to branches. We compute the concentration distributions inside the branches. The originality of the problem is that the geometry of the plant is not a priori given. It evolves in time depending on the concentrations of plant hormones found as a solution of the problem. New branches appear in the process of plant growth. The algorithm is adapted to an arbitrary plant structure and an arbitrary number of branches.

**Keywords:** Plants, branching, variable structure, computational algorithm.

## 1 Introduction

Plant modelling attract much attention both from the point of view of understanding of fundamental biological mechanisms and for practical purposes related crop optimization. Various approaches are developed (see [1], [2]). In this paper we describe a numerical algorithm developed in order to model plant growth and structure formation on the basis of realistic biological mechanism suggested in our previous works [3], [4]. Plants are considered here as interconnected one-dimensional intervals. Each interval corresponds to a branch. There are branches of different generations. There exists only one branch of the first generation. It corresponds to the trunk. One its end is fixed and corresponds to the root, another one moves with some speed. If we introduce a space variable $x$ along this interval, then $x = 0$ corresponds to its fixed end and $x = L(t)$ to the moving boundary. We do not consider the root system and take it into account only by means of the flux of nutrients passing though the fixed boundary. The moving boundary corresponds to the apical meristem, a narrow layer of cells at the very end of the growing branch. These cells divide and determine the plant growth. The cells along the trunk differentiate and do not divide anymore.

We next introduce the concentration of nutrients $C(x,t)$, which depends on the space variable $x$ and on time $t$, and the concentration of growth and mitosis factor $R(t)$, which depends only on time. Nutrient are supplied trough the fixed boundary of the interval and consumed at its moving boundary. Growth and

M.L. Gavrilova et al. (Eds.): Trans. on Comput. Sci. VIII, LNCS 6260, pp. 87–102, 2010.
© Springer-Verlag Berlin Heidelberg 2010

mitosis factor is located in the apex. It is produced there and determines the proliferation rate of the cells, that is the rate of branch growth. Thus, we obtain a free boundary problem where the speed of the boundary is determined by the variable $R(t)$ defined at this boundary and by the variable $C(x,t)$ defined in the whole interval. Precise formulation of the problem will be given in the next section. Biological background of this study is presented in the previous works [3], [4].

Branches of the second generation are also straight intervals with one endpoint inside the first branch (trunk) and another endpoint, which is free. It corresponds to the apex of this new branch and determines its growth. The number of branches of the second generation can be arbitrary, as well as their starting points. These are not fixed parameters of the problem.

In order to describe how branches of the second and of other generations appear, we briefly recall some biological facts. Appearance of new buds is determined by the concentrations of plants hormones auxin and cytokinin. They are produced by the plant itself and redistributed by ascending or descending fluxes. The most recent investigations reveal that auxin efflux carrier PIN1 protein plays the central role in this regulation. The expression of this protein is considered as a key factor for formation of plant organs [5]-[10], [11]-[17]. On the other hand, regulation of the PIN genes is itself under the auxin control (via PLETHORA gene feedback loop [13]). Therefore, we have an auxin induced auxin efflux which can be modelled with or without the intermediate PIN protein. However, auxin by itself cannot initiate cell proliferation. It can happen only in the presence of cytokinin, another plant hormone (actually a group of hormones), the main role of which is the regulation of cell proliferation. Therefore it can be an interplay of these two hormones that causes formation of a new bud.

Thus, if these concentrations take on some given values at some space point, then a new bud appears there. From the algorithmic point of view, buds represent very short branches with associated values of the concentration of mitosis and growth factor. If this value is low, then the bud remains dormant. If it becomes sufficiently high, then the bud gives a new growing branch. Thus, we introduce two new variables $A(x,t)$ and $K(x,t)$ which describe the distributions of auxin and cytokinin. They are defined inside all branches, as well as the concentration of nutrients $C(x,t)$. The space variable $x$ here is proper to each branch. A more precise description of the model is given below.

We summarize the model of plant growth as follows. Plant is represented as a number of connected and growing straight intervals. Their appearance, location and the rate of growth are determined by concentrations of nutrients, hormones and mitosis and growth factors defined either inside the branches or at their boundaries. The plant architecture, that is the number of intervals and their locations are not a priori given. From the mathematical point of view, we consider several free boundary problems whose solutions influence each other. The number of these problems and their relation to each other evolve in time. This is a new type of free boundary problems which we call one-dimensional problems with branching. The numerical algorithm should be able to describe any possible plant architecture. The main purpose of this work is to present this algorithm (Section 3). We illustrate its application to the study of branching patterns in Section 4.

## 2   Model

### 2.1   Without Branching

We consider in this work one-dimensional model justified if the length (or height) $L$ of the plant is essentially greater than the diameter of its trunk. Hence we consider the interval $0 \leq x \leq L(t)$ with the length depending on time. The left endpoint $x = 0$ corresponds to the root. Its role is to provide the flux of nutrients taken into account through the boundary condition. We do not model the root growth here in order to simplify the problem. Therefore the left boundary is fixed. The right endpoint, $x = L(t)$ corresponds to the apex. Its width is much less than that of the plant. We suppose in the model that it is a mathematical point. The value $L(t)$ increases over time. According to the assumption above, the growth rate is determined by the concentration of metabolites at $x = L(t)$, which we denote by $R$. Thus

$$\frac{dL}{dt} = f(R). \tag{2.1}$$

The function $f(R)$ will be specified below.

We recall that the interval $0 < x < L(t)$ corresponds to differentiated cells that conduct nutrients from the root to the apex. We suppose that they are in a liquid solution. Denote by $C$ their concentration, which is a function of $x$ and $t$. Its evolution is described by the diffusion-advection equation

$$\frac{\partial C}{\partial t} + u\frac{\partial C}{\partial x} = d\frac{\partial^2 C}{\partial x^2}. \tag{2.2}$$

Here $u$ is the velocity of the fluid, and $d$ is the diffusion coefficient. Assuming that the fluid is incompressible and fills the xylem uniformly (the part of the plant tissue conducting nutrients from below to above and located inside the cambium layer), we obtain

$$u = \frac{dL}{dt}.$$

We complete equation (2.2) by setting the boundary conditions

$$x = 0 : C = 1, \quad x = L(t) : d\frac{\partial C}{\partial x} = -g(R)C. \tag{2.3}$$

The second boundary condition shows that the flux of nutrients from the main body of the plant to the meristem is proportional to the concentration $C(L, t)$. This is a conventional relation for mass exchange at the boundary, Robin boundary conditions. The factor $g(R)$ shows that this flux can be regulated by proliferating cells. We discuss this assumption as well as the form of the function $g(R)$ below.

We now derive the equation describing the evolution of $R$. At this point we need to return to the model in which the width of the meristem is finite. We denote it by $h$. Then we have

$$h\frac{dR}{dt} = g(R)C - \sigma R. \qquad (2.4)$$

The first term in the right-hand side of this equation describes production of the GM-factor $R$ in the meristem. The second term corresponds to its consumption.

System of equations (2.1)-(2.4) is a generic one-dimensional model of plant growth based on: a) "continuous medium" assumptions of mass conservation (for $C + R$) and of the proportionality of the flux $\partial C/\partial x$ at the boundary to the value of $C$; and b) a "biological" assumption that there is a chemical species $R$, the plant growth and mitosis factor, which is produced in the meristem and which determines the plant growth.

We note that the conservation of mass in the case $\sigma = 0$ implies that the term $g(R)C$ enters both the boundary condition and equation (2.4). Therefore, the assumption that the rate of the plant growth factor production depends on its concentration $R$ makes the boundary condition depend on it also. We will see below that properties of the function $g$ can be crucial for plant growth. In particular, if it is constant (the production rate is not auto-catalytic), we will not be able to describe the essential difference in plant sizes.

We now specify the form of the functions $f$ and $g$. We will consider $f$ as a piece-wise constant function equal to 0 if $R$ is less than a critical value $R_f$ and equal to some positive constant $f_0$ if $R$ is greater than $R_f$ (Figure 1a). This means that growth begins if the concentration of the plant growth factor exceeds some critical value.

The production of the growth factor $R$ is assumed to be auto-catalytic. To simplify the model, we consider a piece-wise linear function $g(R)$ (Figure 1b). In some cases we also consider smooth functions $f$ and $g$. These assumptions are consistent with plant morphogenesis. They are discussed in [3], [4] in more detail.

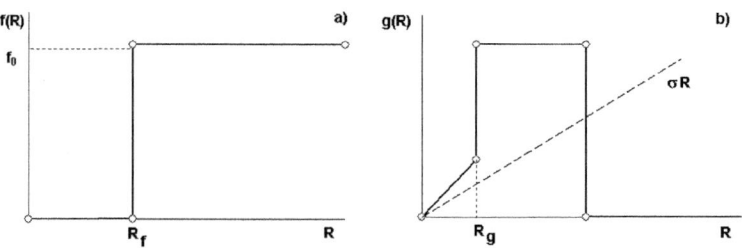

**Fig. 1.** Functions $f$ and $g$

## 2.2   With Branching

The concentrations of nutrients $C$, and of hormones $A$ and $K$ are described by the diffusion equations with convective terms:

$$\frac{\partial C}{\partial t} + V \frac{\partial C}{\partial x} = d_C \frac{\partial^2 C}{\partial x^2}, \tag{2.5}$$

$$\frac{\partial K}{\partial t} - V_K \frac{\partial K}{\partial x} = d_K \frac{\partial^2 K}{\partial x^2} - \mu K, \tag{2.6}$$

$$\frac{\partial A}{\partial t} - V_A \frac{\partial A}{\partial x} = d_A \frac{\partial^2 A}{\partial x^2} - \mu A. \tag{2.7}$$

The convective speed $V_A$ in equations (2.6) and (2.7) can be different in comparison with equation (2.5). It corresponds to transport in the phloem in the direction from top (meristem) to bottom (root). The speed $V$ in the first equation is determined as the speed of growth:

$$\frac{dL}{dt} = V, \quad L\Big|_{t=0} = L_0, \quad h\frac{dR}{dt} = Cg(R) - \sigma R, \quad R\Big|_{t=0} = R_0, \quad V = f(R) \tag{2.8}$$

Here $d_C$, $d_K$, $d_A$ and $\mu$ are parameters; the space variable $x$ is defined independently for each branch.

The boundary conditions for $C$ are the same as in the case without branching:

$$C\Big|_{x=0} = C_0, \quad d_C \frac{\partial C}{\partial x} + Cg(R)\Big|_{x=L} = 0. \tag{2.9}$$

The boundary conditions for $K$ describe its possible production in the root, and its production in the meristem with rate proportional to the rate of growth:

$$K\Big|_{x=0} = K_0, \quad d_K \frac{\partial K}{\partial x}\Big|_{x=L} = \varepsilon V. \tag{2.10}$$

Finally, the boundary conditions for $A$ are similar, except that the boundary condition at $x = 0$ takes into account that this horomone can be transported from the stem to the root:

$$\frac{\partial A}{\partial x} - \beta A\Big|_{x=0} = 0, \quad d_A \frac{\partial A}{\partial x}\Big|_{x=L} = \varepsilon V. \tag{2.11}$$

We define next the branching conditions. A new branch appears at $x = x_0$ and $t = t_0$ if

$$A(x_0, t_0) = A_b, \quad K(x_0, t_0) = K_b, \tag{2.12}$$

where $A_b$ and $K_b$ are some given values. Appearance of a new branch means that there is an additional interval connected to the previous one at its point $x_0$. The variables $C_n$, $A_n$, $K_n$ and $R_n$ are described at the new interval by the same equations as above. Here the subscript $n$ determines the number of the branch. We should complete the formulation by the initial value of the concentration

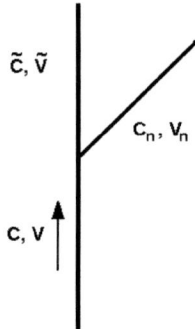

**Fig. 2.** Schematic representation of convective and diffusion fluxes in 1D plant with branching

$R_n = R_n(t_0)$. It cannot be found as a solution of the problem but should be considered as a parameter.

There are additional branching conditions for the concentrations

$$C_n(0,t) = C(x_0,t), \quad A_n(0,t) = A(x_0,t), \quad K_n(0,t) = K(x_0,t),$$

which means that the concentrations are continuous at branching points. We also need some conditions on the fluxes to provide the conservation of mass. Under the notations shown in Figure 2, we have

$$SV = \tilde{S}\tilde{V} + S_n V_n, \quad S\frac{\partial C}{\partial x} = \tilde{S}\frac{\partial \tilde{C}}{\partial x} + S_n\frac{\partial C_n}{\partial x}.$$

Here $S, \tilde{S}$, and $S_n$ are parameters determined by the cross section area of the corresponding branch. If all branches have the same cross section, then $S = \tilde{S} = S_n = 1$. If the cross section areas are conserved and narrow branches have the same diameter, then $S = 1$, and $\tilde{S} = S_n = 1/2$.

In this work we restrict ourselves to the case where all branches have the same cross section. Otherwise, the diameter of each branch should be considered as a function of time. The diameter would depend on fluxes of metabolites coming through the branch to the root. On the other hand, the fluxes of nutrients going through the branch from the root would depend on its cross section. We obtain a very complex time-dependent problem with many feedbacks. This will essentially complicate the understanding of the mechanism of growth and the interpretation of the results.

As is discussed in the previous section, bud formation is accompanied by production of $A$ and $K$. The angle of the new branches with respect to the previous one is given as a parameter. It is not related to light and photosynthesis which are not considered in this model.

# 3    Computational Algorithm

In this section we discuss the numerical algorithm for the one-dimensional problem with branching. The model is based on the reaction-diffusion equations, which can be considered in the following form:

$$\frac{\partial \varphi}{\partial t} + u\frac{\partial \varphi}{\partial x} = d\frac{\partial^2 \varphi}{\partial x^2} - \mu\varphi, \tag{3.13}$$

where $\varphi$ corresponds to the unknown variable (e.g. $C$, $K$), $x$ is the space variable defined on each particular branch. Let us introduce a finite-difference mesh along any branch with the nodal coordinates $x_i$ ($i = 1, \ldots, I$) (Figure 3a).

a)

b)

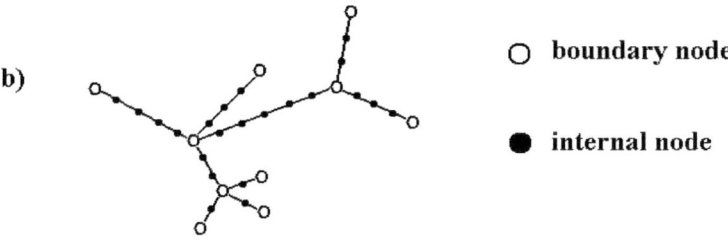

○ **boundary node**

● **internal node**

**Fig. 3.** Numerical mesh with branching

Let the nodes $x_i$ be located uniformly with distance $\delta x$ between them. We will denote the values of $\varphi$ in the nodes by $\varphi_i$. Consider an explicit upwind approximation for equation (3.13) first in the interior nodes ($i = 2, \ldots, I - 1$)

$$\frac{\varphi_i^n - \varphi_i^{n-1}}{\delta t} + (|u| + u)\frac{\varphi_{i+1}^n - \varphi_i^n}{2\delta x} + (|u| - u)\frac{\varphi_{i-1}^n - \varphi_i^n}{2\delta x}$$

$$= d\frac{\varphi_{i+1}^n - 2\varphi_i^n + \varphi_{i-1}^n}{(\delta x)^2} - \mu\varphi_i^n, \tag{3.14}$$

where $\delta t$ is the time step and $\delta x$ the space step. The superscript $n$ corresponds to the time step. To simplify the notations, in what follows we will omit $n$, that is, we will write $\varphi_i$ instead of $\varphi_i^n$, and will keep the notation $\varphi_i^{n-1}$. Let us write equation (3.14) in the form

$$a_i\varphi_{i-1} - c_i\varphi_i + b_i\varphi_{i+1} = -f_i, \quad i = 2, \ldots, I - 1, \tag{3.15}$$

where

$$a_i = \frac{|u| - u}{2\delta x} + \frac{d}{(\delta x)^2}, \quad b_i = \frac{|u| + u}{2\delta x} + \frac{d}{(\delta x)^2},$$

$$c_i = \frac{1}{\delta t} + \frac{|u|}{\delta x} + \frac{2d}{(\delta x)^2} + \mu, \quad f_i = \frac{\varphi_i^{n-1}}{\delta t}. \tag{3.16}$$

Consider now the approximation of the boundary conditions. If the ends of the branch correspond to the root ($x = 0$) or to the growing part (Figure 4a), then the approximation of the boundary conditions here yields the expressions:

$$- c_1 \varphi_1 + b_1 \varphi_2 = -f_1, \tag{3.17}$$

and

$$a_I \varphi_{I-1} - c_I \varphi_I = -f_I. \tag{3.18}$$

Specific expressions for $a_1, b_1, \ldots, f_I$ in the case of some particular boundary conditions can be given in a similar way.

In the case of the growth of a single branch, the numerical scheme is completely described by system (3.15), (3.17), and (3.18). The matrix of this system is tri-diagonal:

$$\begin{pmatrix} -c_1 & b_1 & 0 & 0 & 0 & 0 & 0 & \cdots & 0 \\ a_2 & -c_2 & b_2 & 0 & 0 & 0 & 0 & \cdots & 0 \\ \cdots & & & & & & & & \\ 0 & \cdots & 0 & a_i & -c_i & b_i & 0 & \cdots & 0 \\ \cdots & & & & & & & & \\ 0 & \cdots & 0 & 0 & 0 & a_{I-1} & -c_{I-1} & b_{I-1} \\ 0 & \cdots & 0 & 0 & 0 & 0 & a_I & -c_I \end{pmatrix}. \tag{3.19}$$

In order to solve a linear system with a tri-diagonal matrix (3.19), we can use the Thomas algorithm.

The situation is more difficult if there are several branches. Moreover, the tree can have an arbitrary structure (see, e.g., Figure 3b). There are nodes from which several branches end or begin. Such nodes will also be called boundary nodes. Each of them will be numbered, and the value of $\varphi$ at the $k$-th node is denoted by $\varphi_k$. The boundary conditions in the conventional sense are not imposed in such nodes. Instead, we should take into account conservation of species and fluxes.

Let there exist $J$ branches leaving the boundary node $k$. Then $\varphi_{j,1} = \varphi_k$ at $j = 1, \ldots, J$. The values of $\varphi$ at the second node of each branch will be denoted by $\varphi_{j,2}$, at the third $\varphi_{j,3}$, and so on. Then the conservation of $\varphi$ at the node $k$ can be written as follows:

$$\sum_{j=1}^{J} \left( d \frac{\varphi_{j,2} - \varphi_k}{\delta x} - \mu \frac{\delta x}{2} \varphi_k - \frac{\delta x (\varphi_k - \varphi_k^{n-1})}{2\delta t} - \frac{|u_j| - u_j}{2}(\varphi_k - \varphi_{j,2}) \right) = 0. \tag{3.20}$$

If there are $J$ branches entering the boundary node $k$, then the finite difference scheme at this node will be:

$$\sum_{j=1}^{J} \left( d \frac{\varphi_{j,I-1} - \varphi_k}{\delta x} - \mu \frac{\delta x}{2} \varphi_k - \frac{\delta x (\varphi_k - \varphi_k^{n-1})}{2\delta t} - \frac{|u_j| + u_j}{2}(\varphi_k - \varphi_{j,I-1}) \right) = 0. \tag{3.21}$$

If for some of the branches the numeration begins at the node $k$ and for some other it ends there, then the finite difference scheme can be represented as a superposition of (3.20) and (3.21).

To summarize, we recall that, in the internal nodes of the mesh, we use a three-point scheme (3.15). In the boundary nodes, we have expressions (3.20) or (3.21) (see Figure 4b) that include $\varphi_k$ and the values $\varphi_{j,I-1}$ or $\varphi_{j,2}$ taken at the nodes nearest to the boundary node from the branches connected to this node. Thus we have a completely implicit approximation of our 1D problem.

We describe here the algorithm for the case in which the domain of the computation has several branches arbitrarily connected between each other. Moreover, the number of branches can change during the computation.

The finite difference scheme for the internal nodes of the $j$-th subsystem has the following form:

$$
\begin{cases}
-c_1\varphi_{k_1} + b_1\varphi_2 = -f_1 + \text{other terms,} \\
a_i\varphi_{i-1} - c_i\varphi_i + b_i\varphi_{i+1} = -f_i, \quad i = 2,\ldots,I-1 \\
a_I\varphi_{k_2} - c_I\varphi_{I-1} = -f_I + \text{other terms.}
\end{cases}
\tag{3.22}
$$

The first and the last equations in system (3.22) are obtained from expressions (3.20) and (3.21) corresponding to $k_1$-th and $k_2$-th boundary nodes. The terms corresponding to the $j$-th branch are written explicitly; for other branches they are included in "other terms".

Instead of the tri-diagonal matrix (3.19) we have here an "almost tri-diagonal" matrix for the $j$-th branch:

$$
\begin{pmatrix}
-c_1 & b_1 & 0 & 0 & 0 & 0 & 0 & \ldots & 0 & \text{other terms} \\
a_2 & -c_2 & b_2 & 0 & 0 & 0 & 0 & \ldots & 0 & \\
\ldots & & & & & & & & & \\
0 & \ldots & 0 & a_i & -c_i & b_i & 0 & \ldots & 0 & \\
\ldots & & & & & & & & & \\
0 & \ldots & 0 & 0 & 0 & a_{I-1} & -c_{I-1} & b_{I-1} & & \\
0 & \ldots & 0 & 0 & 0 & 0 & a_I & -c_I & \text{other terms}
\end{pmatrix}.
\tag{3.23}
$$

Step 1. Transformation of matrix (3.23) to an "almost two-diagonal" form. Let us write the corresponding system (3.22) in the form

$$
\begin{cases}
-c_1\varphi_{k_1} + b_1\varphi_2 = -f_1 + \text{other terms,} \\
\beta_i\varphi_{k_1} - \varphi_i + \alpha_i\varphi_{i+1} = -\gamma_i, \quad i = 2,\ldots,I-1, \\
a_I\varphi_{k_2} - c_I\varphi_{I-1} = -f_I + \text{other terms,}
\end{cases}
\tag{3.24}
$$

where

$$
\alpha_i = \frac{b_i}{c_i - \alpha_{i-1}a_i}, \quad \beta_i = \frac{a_i\beta_i}{c_i - \alpha_{i-1}a_i},
$$

$$\gamma_i = \frac{a_i \gamma_{i-1} + f_i}{c_i - \alpha_{i-1} a_i}, \quad i = 3, \ldots, I-1 \tag{3.25}$$

and

$$\alpha_2 = \frac{b_2}{c_2}, \quad \beta_2 = \frac{a_2}{c_2}, \quad \gamma_2 = \frac{f_2}{c_2}. \tag{3.26}$$

Thus, matrix (3.23) is reduced to the following one:

$$\begin{pmatrix} -c_1 & b_1 & 0 & 0 & 0 & 0 & 0 & \ldots & 0 & \text{other terms} \\ \beta_2 & -1 & \alpha_2 & 0 & 0 & 0 & 0 & \ldots & 0 & \\ & \ldots & & & & & & & & \\ \beta_i & \ldots & 0 & 0 & -1 & \alpha_i & 0 & \ldots & 0 & \\ & \ldots & & & & & & & & \\ \beta_{I-1} & \ldots & 0 & 0 & 0 & 0 & -1 & \alpha_{I-1} & \\ 0 & \ldots & 0 & 0 & 0 & 0 & a_I & -c_I & \text{other terms} \end{pmatrix}. \tag{3.27}$$

Step 2. Transformation of the matrix of system (3.24) to an almost diagonal form. Let us write the internal part of system (3.24) in the form:

$$\begin{cases} -c_1 \varphi_{k_1} + b_1 \varphi_2 = -f_1 + \text{other terms}, \\ p_i \varphi_{k_1} - \varphi_i + q_i \varphi_{k_2} = -s_i, \quad i = 2, \ldots, I-1, \\ a_I \varphi_{k_2} - c_I \varphi_{I-1} = -f_I + \text{other terms}, \end{cases} \tag{3.28}$$

where

$$p_i = \alpha_i p_{i+1} + \gamma_i, \quad q_i = \alpha_i q_{i+1},$$
$$s_i = \alpha_i s_{i+1} + \beta_i, \quad i = I-3, \ldots, 2 \tag{3.29}$$

and

$$p_{I-2} = \gamma_{I-2}, \quad q_{I-2} = \alpha_{I-2}, \quad s_{I-2} = \beta_{I-2}. \tag{3.30}$$

Now matrix (3.27) is reduced to the following one:

$$\begin{pmatrix} -c_1 & b_1 & 0 & 0 & 0 & 0 & 0 & \ldots & 0 & \text{other terms} \\ p_2 & -1 & 0 & 0 & 0 & 0 & 0 & \ldots & q_2 & \\ & \ldots & & & & & & & & \\ p_i & \ldots & 0 & 0 & -1 & 0 & 0 & \ldots & q_i & \\ & \ldots & & & & & & & & \\ p_{I-1} & \ldots & 0 & 0 & 0 & 0 & -1 & q_{I-1} & \\ 0 & \ldots & 0 & 0 & 0 & 0 & a_I & -c_I & \text{other terms} \end{pmatrix}. \tag{3.31}$$

Step 3. Determination of the new values of $\varphi$ at the boundary nodes. For any branch we substitute the second equation ($i = 2$) from (3.28) into the first one, and the $(I-1)$-th equation to the last one. We obtain

$$(b_{k_1} p_2 - c_{k_1}) \varphi_{k_1} + b_{k_1} q_2 \varphi_{k_2} = -b_{k_1} s_2 - f_{k_1} + \text{other terms},$$

$$(a_{k_2} - c_{k_2} q_{I-1}) \varphi_{k_2} - c_{k_2} p_{I-1} \varphi_{k_1} = c_{k_2} s_{I-1} - f_{k_2} + \text{other terms.} \qquad (3.32)$$

Let us consider the first and the last equations in (3.32). We see that only the boundary nodes enter these equations. "Other terms" now also contain only the corresponding boundary nodes. Therefore, the boundary nodes form a complete system which allows us to find the unknowns $\varphi$ in all boundary nodes.

Step 4. Determination of new values in the internal nodes. When the values of $\varphi$ are found in all boundary nodes, we can obtain new values of $\varphi$ in the internal nodes for all branches using the corresponding internal equations from system (3.28).

This algorithm is applicable for an arbitrary connection of subsystems. We will finish this section with a short discussion of the sensitivity of the results with respect to numerical discretization. Convergence of the numerical results when we decrease the time and the space steps occurs when we consider finite and not very big time intervals. The situation is much more complex if we consider longtime behavior. Figure 4 shows final (stationary) solutions for different space steps. The absence of convergence of the results can be related to structural instability specific for this model and important for growth of biological organisms (see the discussion in [3]).

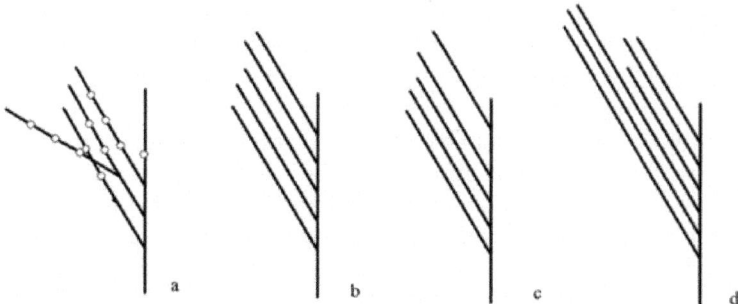

**Fig. 4.** The final (stationary) form of the plant for different space steps: for each figure, the step is half of the previous one

## 4 Branching Pattern

A typical example of plant growth in the 1D model with branching is shown in Figure 5. In the beginning of the evolution there is a single branch which grows with an approximately constant speed until the growth period is finished. The apical meristem is located at the upper end of the interval. The plant hormones auxin and cytokinin are produced there and are transported along the whole branch. It can be diffusive or convective transport. Cytokinin can also be produced in the roots. This is taken into account through the boundary condition at $x = 0$.

If at some point of the branch the concentrations of auxin and of cytokinin take on some prescribed values, then a new bud appears. In the simulation shown in Figure 5, there are five buds that appear one after another at an approximately equal distance. Each of the buds contains its own apical meristem with some value of the GM-factor $R$. When a new bud appears, the initial value $R_0$ is prescribed. It can be some given constant or it can depend on some factors (on plant hormones or on the value of $R$ in the apical meristem of the main growing branch). After that, the value of $R_i$ in the bud evolves according to the same equation (see (2.4)).

When $R_i$ becomes larger than $R_f$, proliferation begins and a new branch starts growing from the bud. As it is shown in [3], growth of branches can be stepwise, that is a branch has periods of growth and of rest during which the speed of its growth equals zero. When the main branch stops growing, under appropriate conditions a new branch can appear from another bud.

As it is discussed above, we consider diffusive and convective fluxes of nutrients. Diffusive flux acts throughout all branches, whereas convective flux is directed to growing branches. This is related to the continuity equation for the incompressible fluid. The routes of convective flux of nutrients is shown in red in Figure 5.

Thus, initiation of growth of new branches is determined by the interplay between the concentrations of nutrients and of the growth factor. In the example presented in Figure 5 there are five buds formed on the main branch. Three of them give branches of the second generation with three new buds on each of them. Branches of the second generation appear one after another when the main branch stops growing. It is interesting to note that one of the buds on the main branch gives a rudimentary branch which stops growing right after it appears. There are also some branches of the third generation.

Similar to the 1D case without branching, the final plant size decreases for larger $h$. In the case with branching, the final length is determined basically by the number of branches and not by their length (Figure 6). There are four generations of branches for $h = 0.0002$, three of them for $h = 0.0005$ and $h = 0.001$, and only two generations for greater values of $h$. A possible explanation of the influence of $h$ on the final length from the point of view of nonlinear dynamics is given in [3].

Plant evolution in time can be influenced by the initial value of the GM-factor concentration in a new bud. We have described it in the case where $R_0 = 0.12$ (Figure 5). In fact, it is the same for all values of $R_0$ between 0.01 and 0.12. Further increase of this parameter changes the plant evolution (Figure 7). When the first bud appears, it does not stay dormant but gives a new branch right away. It grows at the same time with the main branch. The difference between the two cases is determined by the behavior of solutions of equation (2.4). As we have already discussed, when a new bud appears, we prescribe it an initial value $R_0$ of the GM-factor concentration. We recall that the concentration of nutrients $C$ is a function of space and time. Its value $C(x_0, t_0)$ at the new bud determines the right-hand side of this equation: it equals zero at $R = 0$ and at two positive values of $R$, it is negative between first two zeros and positive

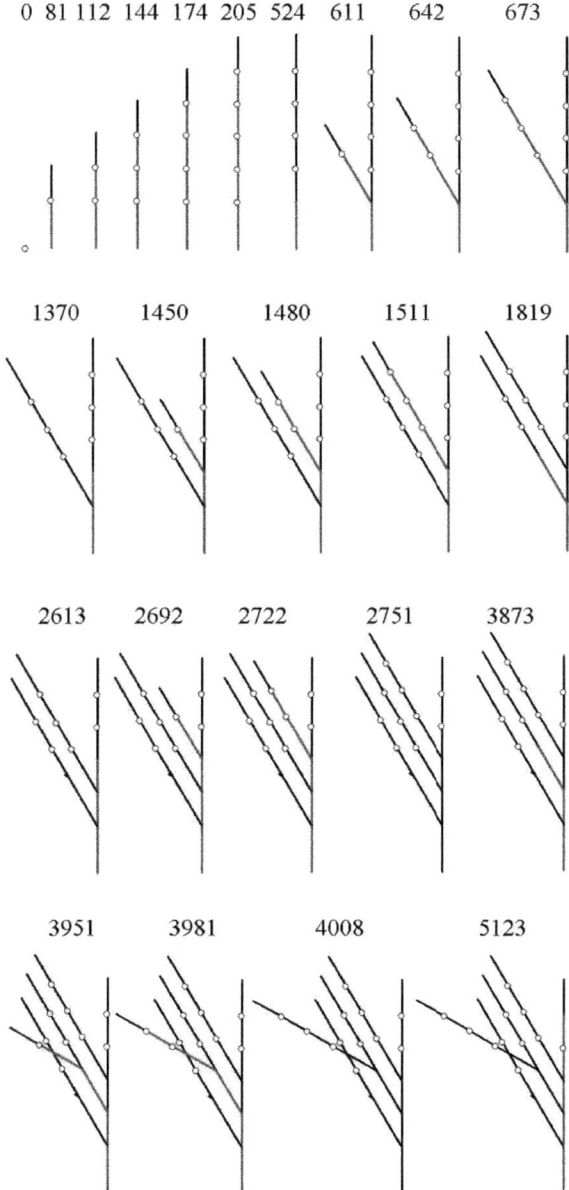

**Fig. 5.** Time evolution of the plant. The figure shows the moments in time when new buds or branches appear. Convective flux of nutrients is shown in red.

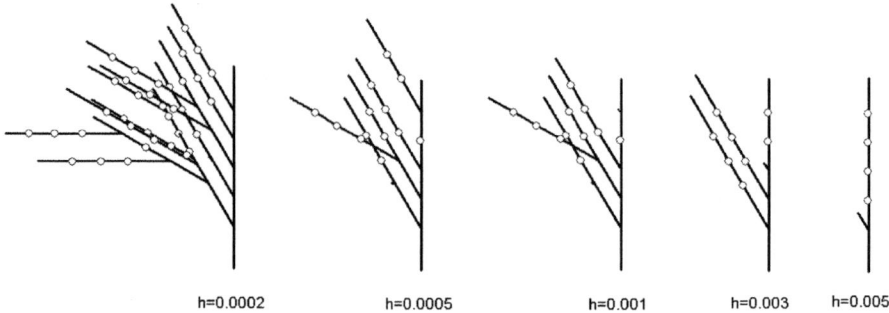

h=0.0002        h=0.0005        h=0.001        h=0.003        h=0.005

**Fig. 6.** Final plant forms for different values of $h$

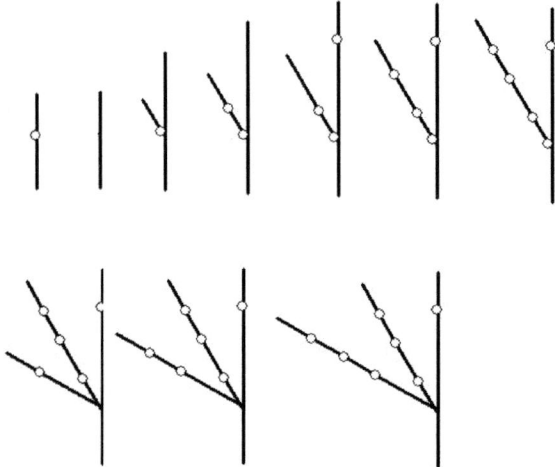

**Fig. 7.** Evolution of the plant structure in time, $h = 0.0005$, $R_0 = 0.13$. Increase of $R_0$ changes completely the plant structure.

between the second two. If the initial value $R_0$ is at the interval of negativity, then the solution rapidly decreases, and the bud remains dormant. When the main branch stops growing, the concentration of nutrients increases. When its value $C(x_0, t)$ at the bud approaches 1, the right-hand side of equation (2.4) becomes positive for small positive values of $R$. The concentration of the GM-factor starts growing. After some time it can reach the critical value which is necessary for the new branch to grow. However, it may happen that another branch will start growing before this one. Then the concentration of nutrient can drop down again, and the concentration of the GM-factor may also decrease.

Formation of new buds and growth of new branches is determined by a complex interaction of plant hormones, nutrients and mitotic factors. The plant structure depends on the values of the parameters. Figure 8 shows the simulations where the buds are double and give symmetric branches. The symmetry of the plant growth is prescribed by the algorithm. Consumption of nutrients

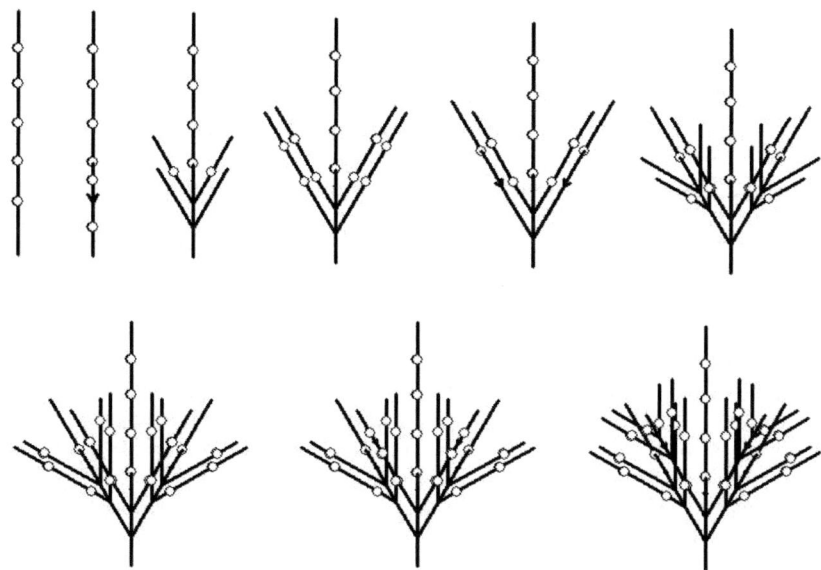

**Fig. 8.** Plant evolution in the case of a double bud and $R_0 = 0.12$

and production of plant hormones are different in this case in comparison with the case without symmetry. Hence the number of new buds and the whole plant organization will also be different.

# References

[1] Krug, H., Liebig, H.-P.: International Symposium on Models for Plant Growth, Environmental Control and Farm Management in Protected Cultivation. ISHS Acta Horticulturae, vol. 248 (1989)

[2] Godin, C., et al.: 4th International Workshop on Functional-Structural Plant Models. Publication UMR AMAP (2004)

[3] Bessonov, N., Volpert, V.: Dynamic Models of Plant Growth. Publibook, Paris (2006)

[4] Bessonov, N., Morozova, N., Volpert, V.: Branching Pattern in Plants. Bull. Math. Biology 70(3), 868–893 (2008)

[5] Heisler, M.G., Ohno, C., Das, P., Sieber, P., Reddy, G.V., Long, J.A., Meyerowitz, E.M.: Patterns of Auxin Transport and Gene Expression during Primordium Development Revealed by Live Imaging of the Arabidopsis Inflorescence Meristem. Current Biology 15, 1899–1911 (2005)

[6] Treml, B.S., Winderl, S., Radykewicz, R., Herz, M., Schweizer, G., Hutzler, P., Glawischnig, E., Ruiz, R.A.: The Gene ENHANCER OF PINOID Controls Cotyledon Development in the Arabidopsis Embryo. Development 139(18), 4063–4074 (2005)

[7] Reinhardt, D.: Regulation of Phyllotaxis. Int. J. Dev. Biol. 49, 539–546 (2005)

[8] Smith, R.S., Guyomarc'h, S., Mandel, T., Reinhardt, D., Kuhlemeier, C., Prusinkiewicz, P.: A Plausible Model of Phyllotaxis. PNAS 103(5), 1301–1306 (2006)

[9] Jonsson, H., Heisler, M.G., Shapiro, B.E., Meyerowitz, E.M., Mjolsness, E.: An Auxin-Driven Polarized Transport Model for Phyllotaxis. PNAS 103(5), 1633–1638 (2006)

[10] Fleming, A.J.: Formation of Primordia and Phyllotaxy. Current Opinion in Plant Biology 8, 53–58 (2005)

[11] Reinhardt, D., Mandel, T., Kuhlemeier, C.: Auxin Regulates the Initiation and Radial Position of Plant Lateral Organs. Plant Cell 12, 507–518 (2000)

[12] Reinhardt, D., Pesce, E.R., Stieger, P., Mandel, T., Baltensperger, K., Bennett, M., Traas, J., Friml, J., Kuhlemeier, C.: Regulation of Phyllotaxis by Polar Auxin Transport. Nature 462, 255–260 (2003)

[13] Blilou, I., Xu, J., Wildwater, M., Willemsen, V., Paponov, I., Friml, J., Heidstra, R., Aida, M., Palme, K., Scheres, B.: The PIN Auxin Efflux Facilitator Network Controls Growth and Patterning in Arabidopsis Roots. Nature 433, 39–44 (2005)

[14] Vernoux, T., Kronenberger, J., Grandjean, O., Laufs, P., Traas, J.: PIN-FORMED 1 Regulates Cell Fate at the Periphery of the Shoot Apical Meristem. Development 127, 5157–5165 (2000)

[15] Galweiler, L., Guan, C., Muller, A., Wisman, E., Mendgen, K., Yephremov, A., Palme, K.: Regulation of Polar Auxin Transport by AtPIN1 in Arabidopsis Vascular Tissue. Science 282, 2226–2230 (1998)

[16] Aida, M., Vernoux, T., Furutani, M., Traas, J., Tasaka, M.: Roles of PIN-FORMED1 and MONOPTEROS in Pattern Formation of the Apical Region of the Arabidopsis Embryo. Development 129, 3965–3974 (2002)

[17] Stieger, P.A., Reinhardt, D., Kuhlemeier, C.: The Auxin Influx Carrier is Essential for Correct Leaf Positioning. Plant J. 32, 509–517 (2002)

# In-Place Linear-Time Algorithms for Euclidean Distance Transform

Tetsuo Asano and Hiroshi Tanaka

School of Information Science, JAIST,
1-1 Asahidai, Nomi, 923-1292, Japan

**Abstract.** Given a binary image, Euclidean distance transform is to compute for each pixel the Euclidean distance to the closest black pixel. This paper presents a linear-time algorithm for Euclidean distance transform without using any extra array. This is an improvement over a known algorithm which uses additional arrays as work storage. An idea to reduce the work space is to utilize the output array as work storage. Implementation results and comparisons with existing algorithms are also included.

## 1 Introduction

There are increasing demands for highly functional peripherals such as printers, scanners, and digital cameras. To achieve intelligence they need sophisticated built-in or embedded softwares. One big difference from ordinary software in computers is little allowance of working space which can be used by the software. In the sense space-efficient algorithms are requested. A number of in-place algorithms have been studied [1,2,3,4,6]. In this paper, we propose another in-place algorithm used for image processing. Especially, we present an algorithm for Euclidean distance transform, which is one of the central problems in image processing. Given a binary image, a distance transform algorithm computes for each pixel how close it is to the closest black pixel.

This paper presents a linear-time in-place algorithm for Euclidean distance transform without using any extra array in addition to a given image matrix. The Euclidean distance transform is to compute a distance map in which each element is the Euclidean distance from the element (pixel) to the closest black element. A number of different algorithms have been proposed and implemented for the Euclidean distance transform [5,7,9,8,10,11]. It has been widely applied in many different problems, especially in medical image analysis [12,13,14].

## 2 Distance Transform

Distance Transform is one of the most important operations in image processing. Given a binary image $G$, our target is to compute a matrix $D$ of the same size such that each element $D(x, y)$ is the distance from each corresponding pixel $(x, y)$ to the closest black pixel (including itself). A simple example of

M.L. Gavrilova et al. (Eds.): Trans. on Comput. Sci. VIII, LNCS 6260, pp. 103–113, 2010.

a binary image is given in Figure 1(a) in which black pixels and white pixels are expressed by black disks and white circles, respectively. The closest black pixel from each white pixel is indicated by an arrow. Figure 1(b) is a matrix of distance transform in which each element is the distance transformation value, the Euclidean distance from the corresponding pixel to the closest black pixel. Note that the distance is 0 for every black pixel.

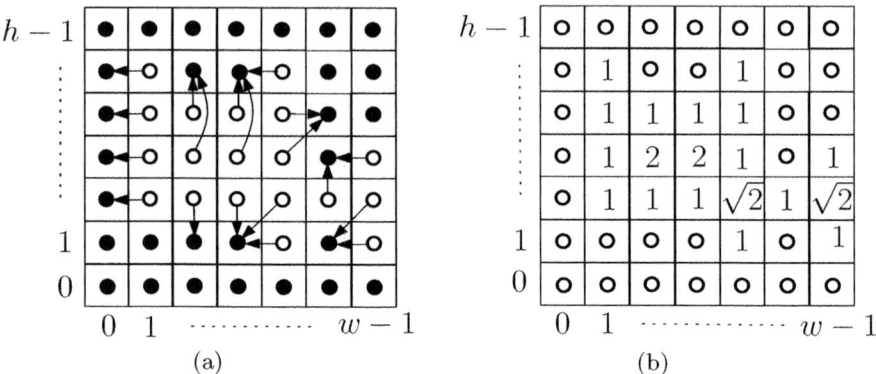

**Fig. 1.** Euclidean distance transform. (a) Closest black pixel (black disk) from each white pixel (white circle) in an binary image $G$. (b) The corresponding distance matrix $D$.

A brute-force algorithm for distance transform is a breadth-first search simulating a wave propagation. Starting from a white pixel, we propagate a wave in the increasing way of the distances from the starting pixel until it encounters some black pixel. It is easy to implement using a queue, but unfortunately, it takes quadratic time in the worst case.

## 3   Known Algorithms

It was open for many years whether the Euclidean distance transform [8] can be computed in linear time or not, but it was finally solved in an affirmative way in 1995 and 1996 by two different groups [10,11]. They are based on different ideas, but both run in linear time in the number of pixels. In this paper we implement the algorithm [11] by Hirata et al. using constant amount of working space.

Their algorithm is described in a pseudo-code as follows.

**Input:** $h \times w$ bit array $G$ for an input binary image with $n = hw$ pixels.
**array:** $h \times w$ array $D$ of squared distances.
**Phase 1:** Vertical Scan
    **for** each $x_c = 0$ **to** $w - 1$
        $d = 0$.
        **for** each $y_c = 1$ **to** $h - 1$

> **if** $G(x_c, y_c) > 0$ **then** $d = d + 1$ **else** $d = 0$.
> $D(x_c, y_c) = d$.

**for** each $x_c = 0$ **to** $w - 1$
  $d = 0$.
  **for** each $y_c = h - 1$ **to** 0
    **if** $G(x_c, y_c) > 0$ **then** $d = d + 1$ **else** $d = 0$.
    $D(x_c, y_c) = \min\{D(x_c, y_c), d\}$.

**Phase 2:** Horizontal Scan
  **for** each $y_c = 0$ **to** $h - 1$
    Initialize a stack.
    Push $(0, D(0, y_c))$ and $(1, D(1, y_c))$ into the stack.
    **for** $x_c = 2$ **to** $w - 1$
      **repeat_forever**{
        $(X_s, Y_s) =$ the rightmost intersection among parabolas in the stack.
        **if** $Y_s \leq (X_s - x_c)^2 + D(x_c, y_c)^2$ **then** exit the loop.
        **else** remove the top element out of the stack.
      }
      Push $(x_c, D(x_c, y_c))$ into the stack.
    **for** $x_c = w - 1$ **down to** 0
      **repeat_forever**{
        $(X_s, Y_s) =$ the rightmost intersection among parabolas in the stack.
        **if** $X_s \leq x_c$ **then** exit the loop.
        **else** remove the top element out of the stack.
      }
      $(x_t, g_t) =$ top_element(stack).
      $D(x_c, y_c) = (x_c - x_t)^2 + g_t^2$.
      // squared distance to the closest black pixel

    The algorithm is divided into two phases. The first phase is to compute the vertical distance to the closest black pixel by two vertical scans, one from bottom to top and the other from top to bottom, while keeping the one-directional distances from black pixel. If we take the smaller distance at each pixel, it is the vertical distance required.

    At the beginning of the second phase the vertical distances are calculated in the matrix $D$. This phase consists of two horizontal scans, one from left to right and the other from right to left. The idea to compute the Euclidean distance is the following. A value $D(x_c, y_c)$ means that the vertically closest black black pixel from the pixel at $(x_c, y_c)$ is either at $(x_c, y_c - D(x_c, y_c))$ or $(x_c, y_c + D(x_c, y_c))$, both in the distance $D(x_c, y_c)$ from $(x_c, y_c)$. It is not important whether the black pixel is above or below it, but the distance $D(x_c, y_c)$ is meaningful. The black pixel at $(x_c, y_c \pm D(x_c, y_c))$ may be the closest black pixel for a different white pixel at $(x_s, y_c)$ in the same row. Its squared distance is given by $(x_c - x_s)^2 + D(x_c, y_c)^2$. So, if we define a parabola $y = (x - x_c)^2 + D(x_c, y_c)^2$ for each pixel $(x_c, y_c)$ in the row $y = y_c$ then the squared distance from a pixel $(x_c, y_c)$ to the closest black pixel is equal to the lower envelope of those parabolas.

It is rather easy to compute the lower envelope of those parabolas since every parabola has the same shape and thus any two intersect at a single point. We construct the lower envelope for the row $y = y_c$ using a stack as follows. The stack initially consists of two elements $(0, D(0, y_c))$ and $(1, D(1, y_c))$, which define parabolas $y = (x - x_c)^2 + D(x, y_c)^2$, $x = 0, 1$. Then, for each $x_c$ from 2 to $n - 1$ we take the top two elements (parabolas) from the stack and compute their intersection $(X_s, Y_s)$ that is the rightmost intersection among parabolas in the stack in the algorithm above. It is indeed easy to compute the intersection of two parabolas, $P_a : y = (x - x_a)^2 + D(x_a, y_c)^2$ and $P_b : y = (x - x_b)^2 + D(x_b, y_c)^2$. If the intersection $(X_s, Y_s)$ lies above the current parabola $P_c : y = (x - x_c)^2 + D(x_c, y_c)^2$, that is, if $Y_s > (X_s - x_c)^2 + D(x_c, y_c)^2$, then the top element (parabola) in the stack never contributes to the lower envelope since the right part of the parabola $P_b$ lies above the current parabola $P_c$ and its left part is above $P_a$. That is why the top element of the stack must be removed. In this way we remove top elements of the stack until the rightmost intersection associated with the stack lies below the current parabola. Then, we push the current parabola into the stack.

Once we have constructed the lower envelope using the stack, we scan the same row of the matrix $D$ from right to left, that is, from $x_c = w - 1$ down to $x_c = 0$. At each pixel $(x_c, y_c)$ in the row, we want to compute the vertical distance to the lower envelope. Since the envelopes in the stack are arranged in the increasing order of the $x$ coordinates of their peaks, we can take out those parabolas in the decreasing order of their $x$-coordinates by popping up the stack. If the parabola just above a pixel $(x_c, y_c)$ is $y = (x - x_t)^2 + g_t^2$ then the vertical distance to the lower envelope is given by $(x_c - x_t)^2 + g_t^2$ (see Figure 2).

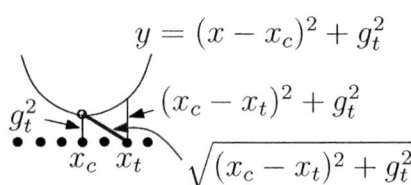

**Fig. 2.** Influence of the white element closest from a black element over other black elements in the same row

**Theorem 1 (Hirata et al. [11]).** *There is a linear-time algorithm for computing the distance transform for a binary image.*

## 4   Constant-Working-Space Algorithm

The algorithm given above is not space-efficient. It uses two working spaces, one for distance matrix $D$ of size $n = h \times w$ and the other for a stack of size $\max\{h, w\}$. It is rather straightforward to reduce the space for the matrix $D$.

It suffices to use the input binary matrix as $D$. Just rename $D$ as $G$, the input image matrix in the algorithm. It causes no problem at all since all the black pixels remain 0 and all the white pixels have positive distances.

It is not so straightforward to reduce the space for the stack. An idea in this paper is to use the distance matrix itself as the stack. When we are computing the distance transform values in the row $y = y_c$, the matrix elements $D(0, y_c), \ldots, D(n - 1, y_c)$ keep the vertical distances to the closest black pixels. First of all we push two elements $(0, D(0, y_c))$ and $(1, D(1, y_c))$ into the stack. Instead of using the stack, we put them at $D(0, y_c)$ and $(0, D(0, y_c))$, that is, we set

$$D(0, y_c) = 0 + w \times D(0, y_c), \text{ and}$$
$$D(1, y_c) = 1 + w \times D(1, y_c).$$

Of course, the original values $D(0, y_c)$ and $D(1, y_c)$ are lost, but they can be restored by dividing the new value $D(0, y_c)$ and $D(1, y_c)$, and then taking their integral parts. We maintain the number of stack elements by a variable, say $t$. Whenever we want to push an element $(x_c, D(x_c, y_c))$ into the stack, we simulate it by setting

$$D(t, y_c) = x_c + w \times D(x_c, y_c),$$
$$t = t + 1 \quad \text{(increment the value of } t\text{)}.$$

Removing the top element out of the stack is easy. Just decrement the value of $t$. Whenever we need an element of a stack, we take the top element $D(t - 1, y_c)$. Then, the corresponding $x$ and $D(x, y_c)$ values can be calculated by

$$x = D(t - 1, y_c) \mod w, \text{ and}$$
$$D(x, y_c) = \lfloor D(t - 1, y_c)/w \rfloor.$$

An important observation is that once we have constructed a stack, the original values of the distance matrix $D$ are not needed anymore. One thing we have to worry about is that whenever we compute the vertical distance at $x = x_c$ we can safely store the value at the corresponding matrix element, $D(x_c, y_c)$. In other words, whenever we store the vertical distance at $D(x_c, y_c)$, the element $D(x_c, y_c)$ must not be a part of the stack. If it is a necessary element of the stack, then it destroys the stack.

It is not so simple to guarantee the safeness. To easily convince ourselves of the safeness we apply an operation called **redundancy removal** to the resulting stack. The operation of redundancy removal is to remove redundant elements from stack. Let $(x_0, g_0), (x_1, g_1), \ldots, (x_{t-1}, g_{t-1})$ be a content of a stack. Each element $(x_i, g_i)$ corresponds to a parabola

$$P_i : y = (x - x_i)^2 + g_i^2, \ i = 0, 1, \ldots, t - 1. \tag{1}$$

Any two consecutive parabolas $P_i$ and $P_{i+1}$ intersect at a single point, which is denoted by $(X_i, Y_i)$. Now, we have some basic observations.

**Observation 1:** $(x_0, x_1, \ldots, x_{t-1})$ is an increasing sequence, that is, $x_0 < x_1 < \cdots < x_{t-1}$.

*Proof.* Immediate from the construction of the stack.

**Observation 2:** A sequence of intersections $(X_0, X_1, \ldots, X_{t-2})$ is also an increasing sequence, that is, $X_0 < X_1 < \cdots < X_{t-2}$.

*Proof.* For contradiction suppose $X_{i-1} > X_i$. By the definition, $(X_{i-1}, Y_{i-1})$ is the intersection of two parabolas $P_{i-1} : y = (x - x_{i-1})^2 + g_{i-1}^2$ and $P_i : y = (x - x_i)^2 + g_i^2$. If the third parabola $P_{i+1} : y = (x - x_{i+1})^2 + g_{i+1}^2$ passes through the intersection $(X_{i-1}, Y_{i-1})$ then we have $X_i = X_{i-1}$, which never happens since it means $P_i$ coincides with $P_{i+1}$. Thus, the inequality $X_{i-1} > X_i$ implies that the point $(X_{i-1}, Y_{i-1})$ lies above the parabola $P_{i+1}$. If it is true, then the parabola $P_i$ must have been removed from the stack when $P_i$ is pushed into the stack. This is a contradiction requested.

Now we define redundant stack elements. As is stated before, the stack gives a sequence of parabolas appearing in the lower envelope. An important notice here is that our goal is not to compute the lower envelope but the vertical distance from each pixel in the current row to the lower envelope. Since each pixel had an integral $x$-coordinate, a stack element is meaningful only if its corresponding parabola appears in the lower envelope at some integral $x$-coordinate. Otherwise, that is, if the interval in which a parabola $P_i$ appears in the lower envelope contains no integral $x$-coordinate, the parabola $P_i$ is useless or **redundant**, which can be safely deleted from the lower envelope (or from the stack) without affecting the calculation of vertical distances.

**Observation 3:** If $X_0 < 0$ then the bottom element $(x_0, g_0)$ in the stack is redundant.

*Proof.* If $X_0 < 0$ holds, then the two parabolas $P_0$ and $P_1$ in the stack intersect at some point to the left of $x = 0$. This means that the parabola $P_0$ lies above $P_1$ in the $x$-interval $[0, w - 1]$. This means that the parabola $P_0$ never appears in the lower envelope in the $x$-interval.

**Observation 4:** If $\lfloor X_{i-1} \rfloor + 1 = \lceil X_i \rceil$ then the element $(x_i, g_i)$ in the stack is redundant.

*Proof.* Recall that $(X_{i-1}, Y_{i-1})$ (resp. $(X_i, Y_i)$) is intersection of two consecutive parabolas $P_{i-1}$ and $P_i$ (resp., $P_i$ and $P_{i+1}$). The equation $\lfloor X_{i-1} \rfloor + 1 = \lceil X_i \rceil$ means that the $x$-coordinates of the two intersections are between two consecutive integers as shown in Figure 3. This means that the middle parabola does not contribute to the lower envelope at any integral $x$-coordinate.

Based on the observations above, we remove all redundant elements from the stack. The first kind of redundant stack elements can be removed by repeatedly applying Observation 3 so that the bottom element gives the lower envelope at $x = 0$. The second kind of redundant stack elements are those which have no

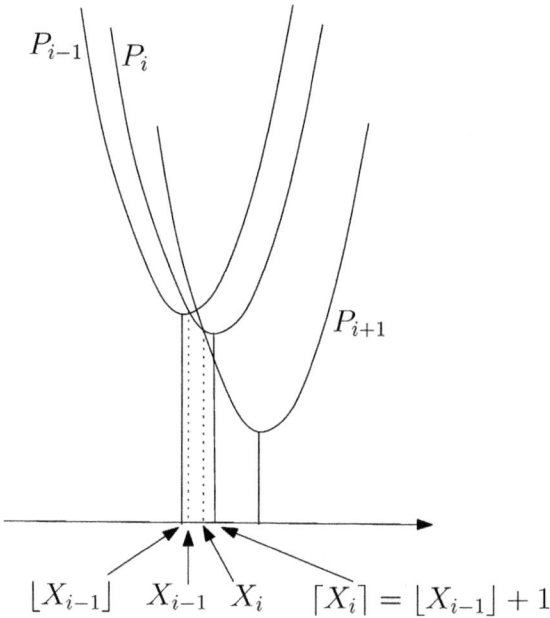

**Fig. 3.** Three parabolas and their intersections

contribution at integral $x$-coordinate. We check two consecutive stack elements and if $\lfloor X_{i-1} \rfloor + 1 = \lceil X_i \rceil$ holds then we remove one stack element as described in Observation 4.

Now the algorithm using only constant amount of working space is given as follows.

**Constant-Working-Space Algorithm for Euclidean Distance Transform**
**Input:** $h \times w$ binary image $G$ with $n = hw$ pixels.
**Phase 1:** Vertical Scan
    **for** each $x_c = 0$ **to** $w - 1$
        $d = 0.$
        **for** each $y_c = 1$ **to** $h - 1$
            **if** $G(x_c, y_c) > 0$ **then** $d = d + 1$ **else** $d = 0.$
            $G(x_c, y_c) = d.$
    **for** each $x_c = 0$ **to** $w - 1$
        $d = 0.$
        **for** each $y_c = h - 1$ **to** $0$
            **if** $G(x_c, y_c) > 0$ **then** $d = d + 1$ **else** $d = 0.$
            $G(x_c, y_c) = \min\{G(x_c, y_c), d\}.$
**Phase 2:** Horizontal Scan
    **for** each $y_c = 0$ **to** $h - 1$
        $G(0, G(0, y_c)) = 0 + w \times G(0, y_c).$
        $G(1, G(1, y_c)) = 1 + w \times G(1, y_c).$

$t = 2$ // Initialize a stack with two elements.
**for** $x_c = 2$ **to** $w - 1$
    **repeat_forever**{
        $(X_{t-2}, Y_{t-2})$ = the rightmost intersection among parabolas in the stack.
        **if** $Y_{t-2} \leq (X_{t-2} - x_c)^2 + G(x_c, y_c)^2$ **then** exit the loop.
        **else** $t = t - 1$ (remove the top element out of the stack).
    }
    $G(t, y_c) = x_c + w \times G(x_c, y_c), \quad t = t + 1$.
    // Push $(x_c, D(x_c, y_c))$ into the stack.
**Phase 3:** Redundancy Removal.
Remove the top element of the stack while it does not contribute to the lower
envelope at $x = 0$.
Remove any stack element whose corresponding parabola has no contribution
to the lower envelope at any integral $x$-coordinate.
**Phase 4:** Final Distance Calculation.
    **for** $x_c = w - 1$ **down to** 0
    **repeat_forever**{
        $(X_{t-2}, Y_{t-2})$ = the rightmost intersection among parabolas in the stack.
        **if** $X_{t-2} \leq x_c$ **then** exit the loop.
        **else** $t = t - 1$ (remove the top element out of the stack).
    }
    $x_a = G(t - 1, y_c) \mod n, \quad g_a = \lfloor G(t - 1, y_c)/w \rfloor$
    // top element of the stack.
    $G(x_c, y_c) = (x_c - x_a)^2 + g_a^2$. // the squared Euclidean distance

**Theorem 2.** *Given a binary image $G$ with $n$ pixels, the algorithm above computes the Euclidean Distance Transform in $O(n)$ time without using any extra array, that is, it outputs a distance matrix of the same size as that of the input image such that each element is the distance from the corresponding pixel to its closest black pixel. Moreover, the largest possible integer stored in the distance matrix during the implementation of the algorithm is bounded by $(n - 1)^2$.*

*Proof.* It is clear that the algorithm does not use any extra array in addition
to an input image array and one for distance map. It is also clear that the
largest possible integer stored in the matrix is at most $(n - 1)^2$, which is greater
than the squared distance between two corner pixels of an input image with $n$
pixels. The correctness of the algorithm follows from the fact that the squared
Euclidean distance from each pixel $p$ to its closest black pixel $q$ is given by the
squared sum of their horizontal distance (defined by their $y$-coordinates) and
their horizontal distance (defined by their $x$-coordinates), which is equal to the
vertical distance between the pixel $p$ and the parabola defined by the pixel $q$.
The algorithm consists of double loops, but in each iteration of the inner loop
the top element of the stack is removed. Since any pixel is popped more than
twice, the total number of iterations in the double loops is bounded by the total
number of pixels, that is, $O(n)$.

In the algorithm above we implement a stack using the input matrix itself. The
output is a matrix with each element being the (squared) Euclidean distance to

its closest black pixel. The largest possible squared Euclidean distance is $(n-1)^2$, which is achieved by a $1 \times n$ array $100 \cdots 00$. Thus, the matrix element must have at least $2 \log_2(n-1)$ bits.

To implement the stack, we put $x$-coordinate $x_c$ and a vertical distance $d$ in a packed manner, that is, $x + w \times d \leq n + w \times h \leq 2n$, which is less than the largest possible value $(n-1)^2$ in the output matrix.

## 5   Bidirectional Distance Transform

Now, consider an extension to a bidirectional Euclidean Distance Transform in which we are requested to compute for each pixel $p$ in a given binary image the distance to the closest pixel with a different value from $p$, that is, distances in both directions from 0 to 1 and from 1 to 0 are required. The bidirectional Euclidean distance transform is rather straightforward if $O(n)$ working space is available. Our goal here is to compute the distance maps using only constant amount of working space. To distinguish distances from 0 to 1 and from 1 to 0, we use signs. For each white pixel the distance is positive, but it is negative for each black pixel.

Idea is the following. In the first two scans we compute vertical distances as before. It is easy to adapt the vertical scan described before to compute bidirectional vertical distances. The resulting distances are stored over the input binary image as before but with signs to distinguish between black and white pixels. The resulting value $d(x, y)$ means existence of opposite-valued pixel in the vertical distance $|d(x, y)|$. If $(x, y)$ is originally white pixel, then $d(x, y)$ is positive. Otherwise, $d(x, y)$ has the minus sign.

Our algorithm is quite similar to the previous one. Suppose we are processing a row $y = y_c$. If the row contains no black pixel, that is, every vertical distance in the row is positive, then we can use the same algorithm as before. So, suppose we have a number of 0-runs and 1-runs in the row. Let $[s, t]$ be an interval of a maximal 1-run in the row. Without loss of generality we assume that

$$p(s - 1, y_c) = 0, p(x, y_c) = 1, \text{ for } x = s, \ldots, t, \text{ and } p(t + 1, y_c) = 0.$$

Here, $p(x, y)$ denotes the original pixel value (0 or 1) at $(x, y)$ in the original binary image. Then, the distance value at any pixel $p(x, y_c)$ in this interval is not affected by vertical distances at pixels in the row in the exterior of the interval. This means that we can compute distance values run by run. An important observation is that a stack fits into the interval of the run (possibly using two more elements corresponding to the black pixels in both ends).

**Theorem 3.** *Given a binary image $G$ with $n$ pixels, we can compute without using any extra array a distance map $d$ in linear time over $G$ such that*
*(1) $d(x, y) > 0$ means that the pixel $(x, y)$ is white pixel and the distance to the closest black pixel is $d(x, y)$, and*
*(2) $d(x, y) < 0$ means that the pixel $(x, y)$ is black pixel and the distance to the closest white pixel is $-d(x, y)$.*

# 6    Concluding Remarks and Future Works

In this paper we have presented a linear-time algorithm for Euclidean Distance Transform which uses only constant amount of working space. We have implemented our algorithm and verified that the running time is almost the same as that of the existing algorithm using stack. An advantage of the proposed algorithm is space-efficiency. It is important for applications in embedded software for visual equipments such as digital cameras.

One of the future works is to extend the idea in this paper to a multiple intensity image. Given an image $G$ with intensities $0, 1, \ldots, g$ for some constant $g \geq 1$, for each pixel $(x, y)$ with intensity $G(x, y) \geq 1$ we want to compute the distance to a closest pixel with intensity less than $G(x, y)$, and also for each pixel $(x, y)$ with intensity $G(x, y) < g$ we want to compute the distance to a closest pixel with intensity greater than $G(x, y)$.

## Acknowledgments

This work of T.Asano was partially supported by the Ministry of Education, Science, Sports and Culture, Grant-in-Aid for Scientific Research on Priority Areas and Scientific Research (B). The author would like to express his sincere thanks to Erik Demaine, Stefan Langerman, Ryuhei Uehara, Mitsuo Motoki, and Masashi Kiyomi for their stimulating discussions.

## References

1. Asano, T., Bitou, S., Motoki, M., Usui, N.: Space-Efficient Algorithm for Image Rotation. IEICE Transactions on Fundamentals of Electronics, Communications and Computer Sciences 91-A(9), 2341–2348 (2008)
2. Asano, T.: Constant-Working-Space Image Scan with a Given Angle. In: Proc. 24th European Workshop on Computational Geometry, Nancy, France, pp. 165–168 (2008)
3. Asano, T.: Constant-work-Space Algorithms: How Fast Can We Solve Problems without Using Any Extra Array? Invited talk at ISAAC 2008, December 2008, p. 1 (2008)
4. Asano, T.: Constant-work-Space Algorithms for Image Processing. In: Nielsen, F. (ed.) Emerging Trends in Visual Computing. LNCS, vol. 5416, pp. 268–283. Springer, Heidelberg (2009)
5. Borgefors, G.: Distance transformations in digital images. Computer Vision, Graphics and Image Processing 34, 344–371 (1994)
6. Chan, T.M., Chen, E.Y.: Multi-Pass Geometric Algorithms. Discrete & Computational Geometry 37(1), 79–102 (2007)
7. Chen, L., Chuang, H.Y.H.: A fast algorithm for Euclidean distance maps of a 2-D binary image. Information Processing Letters 5(1), 25–29 (1994)
8. Fabbri, R., da F. Costa, F., Torelli, J.C., Bruno, O.M.: 2D Euclidean Distance Transform Algorithms: A Comparative Survey. ACM Computing Surveys 40(1), 2:1–2:44 (2008)

9. Gabrilova, M.L., Alsuwaiyel, M.: Computing the Euclidean distance transform. Journal of Supercomputing 25, 177–185 (2003)
10. Breu, H., Gil, J., Kirkpatrick, D., Werman, M.: Linear Time Euclidean Distance Algorithms. IEEE Trans. on Pattern Analysis and Machine Intelligence 17(5), 529–533 (1995)
11. Hirata, T.: A unified linear-time algorithm for computing distance maps. Information Processing Letters 58(3), 129–133 (1996)
12. Klein, F., Kübler, O.: Euclidean distance transformations and model guided image interpretation. Pattern Recognition Letters 5, 19–20 (1987)
13. Paglieroni, D.W.: Distance Transforms. Computer Vision, Graphics and Image Processing: Graphical Models and Image Processing 54, 56–74 (1992)
14. Rosenfeld, A., Kak, A.C.: Digital Picture Processing, 2nd edn. Academic Press, New York (1978)

# A Foundation of Demand-Side Resource Management in Distributed Systems

Shrisha Rao

International Institute of Information Technology - Bangalore
shrao@alumni.cmu.edu

**Abstract.** The theoretical problems of demand-side management are examined in without regard to the type of resource whose demand is to be managed, and the Maximum Demand problem is identified and addressed in a system consisting of independent processes that consume the resource. It is shown that the Maximum Demand problem generalizes the Santa Fe Bar Problem of game theory, and the basic properties of Maintenance and Strong and Weak Recovery that are desirable of systems of processes where demand management is practiced are defined. Basic algorithms are given for solving the Maximum Demand problem in a general context, and in a system where the processes have priorities.

**Keywords:** Maximum demand, Santa Fe Bar Problem, maintenance, recovery.

## 1 Introduction

One of the issues faced by designers and users of large systems, which are almost as a matter of definition distributed, is the need for resource management. Though resource management is traditionally a concern for large systems such as cities and industries, such a need is true even of large computing systems. For instance, though in the age of personal computing the power consumed was not traditionally regarded as a great burden, it is so now at two levels: (i) smaller mobile and embedded devices (and distributed systems consisting of or including such) face severe constraints on account of limited battery power; and (ii) some large server farms, data centers, and other distributed computing systems now consume power costing millions [16], and thus need to be managed [17] just as with other large industrial concerns. In the case of large systems, in addition to the power consumed by the computing devices themselves (all of which gets dissipated in the form of heat), the power consumed by the inevitable air-conditioning systems is also significant, and may cost an additional dollar for each dollar spent on powering up the computing systems themselves [16].

In addition to large computing systems, with the proliferation of embedded control in industrial equipment and the like, the field of distributed computing has been given a vast new array of problems in the area of demand-side management in large systems. The relevance and applicability of this matter are already well-established, given the widespread awareness of the need for energy efficiency, as well as the imminent threat of global warming.

M.L. Gavrilova et al. (Eds.): Trans. on Comput. Sci. VIII, LNCS 6260, pp. 114–126, 2010.
© Springer-Verlag Berlin Heidelberg 2010

In this paper, the concerns addressed are mainly applicable to large systems, including mobile or embedded ones, that draw significant amounts of some resource such as electrical power. However, some of the suggested insights may hold true for smaller systems (e.g., those consisting of battery-operated devices) as well, though this needs to be explored further.

There of course are many engineering studies of demand-side management (called demand response in the power engineering community), which address issues such as overall efficiency [18], possible strategies to be adopted by consumers of power [6], and prevention of failures in the power supply system through demand management [12]. (These are of course only indicative examples, and many more such exist.) However, these works do not address the computing problems that arise when any planned solutions are to be implemented. The purpose of this paper is to demonstrate that there exists a class of problems arising in the realm of demand-side resource management of large systems that can be evaluated in the abstract using the tools of distributed computing.

A somewhat naïve consideration of resource management or optimization in a system would be to just associate it with a reduction in the resource consumed by the system, or equivalently, with a maximization of the quantum of output delivered by the system per unit of resource consumed. However, based on analogies from engineering, it seems reasonable that the following (of which we look at the first in this paper) all be considered as possible demand-side management problems.

(1) Maximum Demand—the rate of resource usage by a distributed system at any given time should be less than or equal to some maximum limit (which itself may change periodically), with any excess subject to a tariff penalty.
(2) Total Demand—the total amount of resource used by a distributed system that performs a given amount of work should be the minimum possible, or conversely, the system should perform the most possible work for a given amount of resource consumed.
(3) Demand Responsiveness—the demand-side management of the system should meet certain standards in terms of quality factors of responsiveness (including, but not limited to, speed of response).
(4) Demand Balance—in cases where a resource is drawn from more than one producer, the demand for it must be balanced in a way that avoids taxing any one producer too greatly.

Though the discussion of demand management in the literature often mentions only electrical power, much of it can be easily applied in other contexts where other consumable resources (e.g., water, fuel) are used in a distributed system. It is therefore not always essential to be very specific about the resource being considered, and it is possible to formulate some general principles and analyses of demand-side management.

The advances in this paper are: (i) formulations of some of the primary demand-management problems, and in particular the characterization of the Maximum Demand problem as a generalization of the Santa Fe Bar Problem;

(ii) characterization of the properties of Maintenance and Recovery that are desirable in systems where the Maximum Demand constraint is sought to be maintained; and (iii) Maximum Demand algorithms for general systems as well as systems where processes have priorities.

The rest of this paper is as follows. The next section describes the basic notation used throughout (other notation is introduced in appropriate contexts later), and uses this to describe in mathematical form the problems that could be addressed. Section 3 takes a better look at one of these, Maximum Demand, with Section 3.1 showing that Maximum Demand is a generalization of the Santa Fe Bar Problem, and Section 3.2 defining the properties of Maintenance and Recovery that are required of self-adaptive distributed systems where Maximum Demand is a concern. Section 3.3 gives a minimal, general Maximum Demand algorithm, and the conditions for Maintenance and Recovery to obtain in it, and Section 3.4 considers how Maximum Demand could be addressed in a system where processes have priorities. Section 4 offers some conclusions and suggestions for further work.

## 2    Notation and Problems

Let there be numerous processes $p_i$ that consume some resource and together constitute the whole system, with the set of processes being denoted as $\mathcal{C}$. Processes always consume a non-negative quantity of the resource (i.e., they are not producers). However, it is possible for a process to consume zero amount of a resource (this may appear unreasonable or counter-intuitive, but in such cases we can conceive of the process as controlling some large piece of equipment, with its own resource consumption being negligible in comparison with that of the equipment it controls).

The set $T$ represents the set of instants of time, and $\mathbb{R}^+$ is the set of positive real numbers, used in this case to stand (in some appropriate unspecified units) for the resource drawn. The *demand function* $\delta : \mathcal{C} \times T \to \mathbb{R}^+ \cup \{0\}$ gives the consumption of a process at a given time, so that $\delta(p_i, t) = r$ means that the resource drawn by process $p_i$ at time $t$ is $r$.

The *maximum resource limit* denoted $M_t$ is the maximum rate of resource consumption allowed at a given time $t$. The *actual resource drawn* denoted $S$ is the current rate at which resource is drawn. These are both on the whole system, not per individual process.

Some of the previously-mentioned problems that need to be considered can now be described as follows.

- The Maximum Demand constraint can be stated, in its most basic form, as the problem of maintaining the condition:

$$\forall t, \ \sum^{i} \delta(p_i, t) \leq M_t. \tag{1}$$

- The Total Demand constraint can be stated as the problem of ensuring that the following holds over a duration of time starting with $t_1$ and ending at $t_2$:

$$\int_{t_1}^{t_2} \sum^{i} \delta(p_i, t)dt \leq D, \qquad (2)$$

where $D$ is the greatest total consumption that is allowed to occur system-wide between times $t_1$ and $t_2$.

- The Demand Balance constraint can be stated as the problem of maintaining the condition:

$$\forall t, \ |\sum^{i} \delta(p_i, t) - \sum^{j} \delta(p_j, t)| \leq a, \qquad (3)$$

with $p_i \in I, p_j \in J, I, J \subseteq \mathcal{C}, I \cap J = \emptyset$, and some $a \geq 0$. Informally, $I$ and $J$ are sets of processes that obtain a resource from different producers.

There can be other characterizations of balance, e.g., if a single process can at once draw from more than one producer, and if different producers have different capacities.

Demand Responsiveness is not characterized here. It can be characterized in various ways depending on what specific aspects of responsiveness one is interested in (e.g., the Strong Recovery property, relation (6) in Section 3.2, can qualify as a characterization, though not as the only one).

# 3    Maximum Demand

In the present paper, we restrict the detailed discussion to only the Maximum Demand problem. Doing this is appropriate from a theoretical as well as a practical perspective: the development of the theory in this case seems foundational for all further work, and the Maximum Demand problem is also, on anecdotal evidence,[1] more important to designers and users of systems than any of the others. It is also a problem that is more readily treated in the manner of classical computing problems.

## 3.1    Maximum Demand and the Santa Fe Bar Problem

The Santa Fe Bar Problem (SFBP), also called the El Farol (Bar) Problem, was conceived by W. Brian Arthur [2], an economist at the Santa Fe Institute, whose description of it is in part as follows.

---

[1] For instance, a power supply utility is generally not unhappy with customers who consume more over a period of time, as this means greater revenue for it; conversely, customers are aware of the financial implications, and tend to make efforts to consume less overall. However, a disconnect between the utility and its customers can and does arise when it comes to maximum demand, e.g., when customers all want to run air conditioners on a hot day, and the utility would rather they did not.

N people decide independently each week whether to go to a bar that offers entertainment on a certain night. For concreteness, let us set N at 100. Space is limited, and the evening is enjoyable if things are not too crowded—specifically, if fewer than 60% of the possible 100 are present. There is no way to tell the numbers coming for sure in advance, therefore a person or agent: goes—deems it worth going—if he expects fewer than 60 to show up, or stays home if he expects more than 60 to go. (There is no need that utility differ much above and below 60.) Choices are un-affected by previous visits; there is no collusion or prior communication among the agents; and the only information available is the numbers who came in past weeks. (The problem was inspired by the bar El Farol in Santa Fe which offers Irish music on Thursday nights; but the reader may recognize it as applying to noontime lunch-room crowding, and to other coordination problems with limits to desired coordination.)

This problem is in some senses similar to the Tragedy of the Commons [10], and has been extensively studied, finding applications in areas such as distributed database replication [7], load balancing in wireless LANs [1], and the emergence of heterogeneity in multi-agent systems [4]. It is an archetype of a repeated non-cooperative game among multiple agents with no coordination, and is well known not to possess a Nash Equilibrium. The game results in the bar going through alternations of overfull and mostly empty, i.e., oscillatory states in the classical sense of Cournot [3], rather than converging to stable behavior [9] (a steady number of patrons visiting the bar), even as the number of occurrences tends to infinity. This happens because of the fundamental contradiction perhaps best captured by the famous Yogi Berra quip, "Nobody goes there any more; it's too crowded." If all agents, who are identical and do not communicate, believe that the bar will be crowded one evening, they choose, with equal high probability, to stay home, resulting in a mostly empty bar; likewise, if they all believe that the bar will not be crowded one evening, they decide with the same high probability to go there, resulting in an overfull bar. Therefore, "rationality precludes learning" [8] in this problem.

The SFBP is more than just mildly interesting for an important reason: the Maximum Demand problem is a generalization of the SFBP. In the relation (1) in Section 2, if $t$ is constrained to take discrete values, and if $\delta(p_i, t) \in \{0, 1\}$, then what results is the Santa Fe Bar Problem.

Therefore, the Maximum Demand problem also shares the characteristics of the SFBP such as oscillatory behavior in the absence of coordination.

A further formal problem related to the SFBP is the Potluck Problem [5], which models the situation with variations in both demand and supply, and can be described with an informal example as follows. There are many agents that have a potluck dinner regularly (e.g., each week). Each of them decides individually how much food to contribute to the dinner. The agents do not communicate among themselves except that every agent knows about the total demand and supply at the dinners in previous weeks. The dinner is enjoyable if there is no *starvation* or *excess* of food. Starvation means that the amount of food brought to the dinner is not sufficient to serve the people in the dinner, while excess means that some food goes waste for being more than sufficient for the dinner. The demand for food by each agent varies according to variable individual appetite each week. An agent decides on how much to contribute depending on the prediction it makes for the demand and supply in that week.

It is clear that the Potluck Problem (with the demand being fixed, and the supply varying) would be similar to the SFBP; alternatively, the SFBP would be equivalent to the Potluck Problem if the capacity of the bar were to also vary week by week.

It is known [5] that a weighted-majority learning approach [14] significantly reduces the oscillations seen in the Potluck Problem, i.e., produces a better parity between demand and supply in the absence of perfect information and communication between producers and consumers.

## 3.2   A Closer Look at Maximum Demand

We can identify two key properties that are sought in self-adaptive systems where maximum demand management is practiced:

**Maintenance** is the property of a system that if the actual resource drawn is less than or equal to the maximum demand at a given time, then it is also less than or equal to the maximum at every future time, provided the allowable maximum does not change. This can be indicated as:

$$\sum^i \delta(p_i, t_1) \le M_t \rightarrow \sum^i \delta(p_i, t_2) \le M_t, \tag{4}$$

where $t_1, t_2 \in T$, $t_2 - t_1 \ge 0$.

**Recovery** is the fault-tolerance property of a system that if the actual demand exceeds the allowable maximum at some time (e.g., due to a transient failure, or due to a sudden decrease in the allowable maximum), then the actual resource drawn will be less than or equal to the allowable maximum at some finite time in the future. This can be further split into two forms, with *weak* recovery being that a system will not take infinitely long to recover, and *strong* recovery being that it will recover within a given finite time $k$.

These can be indicated as:

(Weak Recovery): for any $t_1 \in T$, there exists a $t_2 \in T$, with $t_2 - t_1 < \infty$, such that

$$\sum^i \delta(p_i, t_1) > M_t \rightarrow \sum^i \delta(p_i, t_2) \le M_t, \tag{5}$$

and:

(Strong Recovery): there exists a $k < \infty$, such that for all $t_1, t_2 \in T$ where $t_2 - t_1 \ge k$,

$$\sum^i \delta(p_i, t_1) > M_t \rightarrow \sum^i \delta(p_i, t_2) \le M_t. \tag{6}$$

The maintenance property indicates that a system does not, on its own, stop obeying the maximum demand constraint, while either recovery property indicates that a system that is not presently compliant with such a constraint will become so in a finite time. Weak Recovery says that a system that does not presently satisfy the maximum demand constraint will do so within a finite time. Strong recovery actually specifies a finite bound $k$ and says that a system that is not presently compliant will be compliant for all future times $k$ or later.

Weak Recovery is obviously required behavior for a system to have any claim of self-adaptive behavior in response to dynamic changes in the maximum resource limit, but practical electric supply systems may require rather stringent Strong Recovery behavior—in such cases, $k$ in (6) may be very small (for instance, an industrial system that consumes more than the maximum limit for over 30 seconds may be subject to a disproportionately large tariff penalty).

### 3.3    Basic Maximum-Demand Algorithm

Consider a greatly simplified algorithm for the Maximum Demand problem. We assume that the maximum resource limit is large enough to let at least some processes get sufficient resource. No process crashes or otherwise fails at any time. All processes in the system are *well-behaved* because they satisfy the following properties while remaining functional:

1. A process needing more resource, when given a constrained value by which it is allowed to increase its resource drawn, restricts its increase to within that value.
2. A process that has a reduced need for resource (perhaps because it has completed some task, or because its working conditions have changed) voluntarily reduces the amount of resource it draws. (This also precludes a process initially drawing more of a resource than it needs.)
3. Similarly, if a process is asked to reduce its intake (in order to facilitate system recovery), it does so voluntarily to the greatest extent it can.

This type of self-governance is not usually considered relevant in computational contexts (e.g., in operating systems), but it is a fair model of the behavior reasonably expected of agents or processes in industrial and other contexts. This is possibly because traditionally, resources in computing systems are often considered non-preemptable, or because standard models of programming and system development often take an all-or-nothing approach wherein a process is not designed to be *gracefully degrading* in the absence of its full quota of desired resources—it either does everything it is supposed to, or else little to nothing.

In the present instance, the following are global variables in the system: $M_t$ and $S$, where $M_t$ is the maximum resource limit, and $S$ is the actual resource drawn. There are three local variables per process $p_i$, with $\delta(p_i, t)$ indicating the resource drawn, $x_i$ indicating the desired change, and $y_i$ indicating a forced reduction.

The following simplified algorithm then suffices for the Maximum Demand problem, starting with system initialization.

**Algorithm 3.1:** (*Max Demand − General*)

(Pseudocode for process $p_i$)

**while** ($\neg$condition(i)) wait;
 /* if entry condition for $p_i$ not true, wait */

**if** ($M_t \geq S$)
 /* if currently drawn is no greater than maximum */

**do** $\begin{cases} \text{(assign } x_i, \text{ such that } x_i \leq M_t - S) \\ \quad /* \text{ fix desired change } */ \\ (\delta(p_i, t) \leftarrow \delta(p_i, t) + x_i) \\ \quad /* \text{ change resource drawn } */ \\ (S \leftarrow S + x_i) \\ \quad /* \text{ increment global variable to reflect change } */ \end{cases}$

 **else**
 /* if currently drawn is more than maximum allowed */

**do** $\begin{cases} \text{(assign } y_i, \text{ such that } 0 \leq y_i \leq S - M_t) \\ \quad /* \text{ fix quantum of reduction } */ \\ (\delta(p_i, t) \leftarrow \delta(p_i, t) - y_i) \\ \quad /* \text{ reduce resource drawn } */ \\ (S \leftarrow S - y_i) \\ \quad /* \text{ decrement global variable to reflect change } */ \end{cases}$

The algorithm works as follows.

A. Each process $p_i$ checks for an entry condition to allow it to execute changes (the entire code except for this check is considered to be a critical section). Such a condition can be implemented using mutex locks, monitors, or similar.[2] If the condition for $p_i$'s entry into the critical section (here called condition(i)) is not true, then $p_i$ has to wait until it becomes true.
B. Once it enters the critical section, $p_i$ compares $M_t$ and $S$.
  - If the former is greater, which means that there is less of the resource being drawn by the system than the maximum resource limit, then it picks an amount $x_i$ not greater than the difference between these two, draws that much more of the resource, and updates the global variable $S$ before exiting the critical section. (Note that this step also suffices for $p_i$ to *reduce* its resource consumption; in that case $x_i$ is negative, which is permissible.)
  - If $S$ is greater than $M_t$, then $p_i$ is obliged to pick an amount $y_i$ that may be 0 or greater, and reduce its consumption by this amount.

---

[2] A "lock-free" or "wait-free" implementation is possible but is not attempted, as doing so would distract much attention away from the main issues, and towards proving the correctness and plausibility of such an implementation. Readers who are averse to critical sections are invited to come up with their own wait-free versions using their favored hardware or software primitives.

It is obvious that this algorithm satisfies the Maintenance property, as no process increases its resource drawn by an amount large enough to cause a true constraint to become false. This simple algorithm as given does not guarantee even the Weak Recovery property, for it is possible that $y_i$ may be 0 always, resulting in no reduction in $S$; but it is trivial to show that the Weak Recovery property holds if $y_i$ is always greater than some $\epsilon > 0$ (a somewhat unreasonable assumption, as it means that every process in the system reduces the resource it draws by more than a certain amount).

The following indicates the necessary and sufficient conditions for Strong Recovery in the system.

**Theorem 1.** *A system that runs Algorithm 3.1 can guarantee Strong Recovery in $k$ if and only if:*

(i) *some number $j > 0$ of processes $p_i$ execute the code in any time interval of length $k$; and*

(ii) *for those $j$ processes, $\sum^j y_i \geq S - M_t$.*

*Proof.* The first part is obvious: if no process executes the code during some period of time $k$, then there must exist times $t_1$ and $t_2$ which are $k$ or more apart, such that:

$$\sum^i \delta(p_i, t_1) > M_t \equiv \text{TRUE} \quad \& \quad \sum^i \delta(p_i, t_2) \leq M_t \equiv \text{FALSE}.$$

Therefore, the condition (6) is false, meaning that Strong Recovery does not hold, and having a non-zero number of processes execute the code is a necessary condition.

For the second part, if the $j$ processes that execute the code in time $k$ together reduce their resource drawn by $\sum^j y_i$, then the value of $S$ is reduced to $S - \sum^j y_i$, which by the relation given is less than $M_t$. Therefore the system is now in compliance with the Maximum Demand constraint, meaning that the second condition is sufficient for Strong Recovery. □

Note that in Algorithm 3.1 it is not necessary for the overall scheduling of processes (or the access of processes to the critical section) to be fair, as in avoiding starvation of some process; Maintenance and Recovery can exist even when fairness is absent.

## 3.4   Maximum-Demand Algorithm Considering Priorities

So far we have said nothing about the priorities of the various processes $p_i$. In practice, however, such distributed processes drawing a resource are almost always distinguished from one another in terms of their priorities. Such distinctions may arise because some processes in the system are inherently more important and perform safety-critical or mission-critical functions, or perhaps because their utilities are not equal or constant. In such a case, the following basic principle needs to be observed.

**Proposition 1.** *For any given amount of resource available, the higher-priority processes have the greater preference, and in any situation where there is a shortfall, the lower-priority processes have the greater preference in any required cutbacks.*

In other words, a higher priority process receives more of a resource than a lower-priority one in times of scarcity, by being asked to sacrifice less when the availability is reduced, and also by being given a greater share when more of the resource becomes available.

This once again is not an issue that is traditionally dealt with in this way in the specific context of computing systems, but it is well within reason in the context of larger systems, where critical functions and processes typically receive a greater share of available resources, and also suffer the least when there have to be cutbacks on account of resource scarcity. This is a concept that is well understood in many non-computing contexts, and is explicated at some length in the context of water resources by Hill [11] and earlier, in a classic monograph on health services, by Klein et al. [13].

Therefore, we need to modify our previous analysis, which took no account of processes' priorities. Specifically, Algorithm 3.1 needs to be modified suitably to take into account the priorities of processes. It is trivial to make such a modification assuming that processes have exactly two priorities, but a general algorithm with many priorities possible can also be given.

In order to favor the higher-priority processes over lower-priority ones in dealing with scarcity, the basic insight applied is that if the maximum resource limit is greater than the actual resource drawn, then processes are given a chance to increase their intake in *decreasing* order of priorities (i.e., with the highest-priority processes going first), whereas if the actual resource drawn is greater than or equal to the maximum resource limit and the system needs to recover, then processes are asked to decrease their intake in *increasing* order of priorities (i.e., with the lowest-priority processes going first).

Formally, let there be a function $\varphi : \mathcal{C} \to \mathbb{R}^+ \cup \{0\}$ that maps processes to priorities, so that $\varphi(p_i) = m$ is a way of saying that the priority of process $p_i$ is $m$. Without loss of generality, let the processes be numbered from 0 to $n - 1$ and ordered in non-increasing order of priority, so that $i > j \to \varphi(p_i) \leq \varphi(p_j)$. As before, there are global variables $M_t$ and $S$, and local variables per process of $\delta(p_i, t)$, $x_i$, and $y_i$. In addition, the controller now has a variable $i$ denoting the index of a process. We use $M_t \updownarrow$ to denote a change (either a decrease or an increase) in the value of $M_t$.

The processes themselves are well behaved, i.e., satisfy the conditions 1–3 given in Section 3.3. However, in this instance, a controller or scheduler is required that would order the processes by priority in giving them entry to the critical section. (If such a controller is not used and processes are allowed to run independently, then priority inversion may occur.) For this basic development, questions such as what would be a reasonable implementation of a (distributed) controller, or how the processes and the controller communicate, are not discussed.

**Algorithm 3.2:** (*Max Demand – Prioritized*)

(Pseudocode for controller)

**while** (TRUE) /* repeat forever */

**do** $\left\{ \begin{array}{l} \textbf{if } (M_t \geq S) \\ \quad /* \text{ currently drawn no greater than max } */ \\ \textbf{for } i \leftarrow 1 \textbf{ to } n-1 \\ \quad \textbf{do} \left\{ \begin{array}{l} (\text{assign } x_i, \text{ such that } x_i \leq M_t - S) \\ /* \text{ fix desired change } */ \\ (\delta(p_i, t) \leftarrow \delta(p_i, t) + x_i) \\ /* \text{ change resource drawn by } p_i* / \\ (S \leftarrow S + x_i) \\ /* \text{ increment global variable } */ \\ \textbf{if } (M_t \updownarrow) \textbf{ break} \\ /* \text{ value of } M_t \text{ changed, quit } \textbf{for} \text{ loop } */ \end{array} \right. \\ \\ \textbf{else} \\ \quad /* \text{ if currently drawn is more than maximum } */ \\ \textbf{for } i \leftarrow n-1 \textbf{ to } 1 \\ \quad \textbf{do} \left\{ \begin{array}{l} (\text{assign } y_i, \text{ such that } 0 \leq y_i \leq S - M_t) \\ /* \text{ fix quantum of reduction } */ \\ (\delta(p_i, t) \leftarrow \delta(p_i, t) - y_i) \\ /* \text{ reduce resource drawn by } p_i \text{ } */ \\ (S \leftarrow S - y_i) \\ /* \text{ decrement global variable } */ \\ \textbf{if } (M_t \updownarrow) \textbf{ break} \\ /* \text{ value of } M_t \text{ changed, quit } \textbf{for} \text{ loop } */ \end{array} \right. \end{array} \right.$

The algorithm works as follows.

A. If $S$ is less than $M_t$, i.e., if less of the resource is being drawn than the maximum resource limit, then the top **for** loop, which schedules processes in decreasing order of priority, is executed.
  - Each process $p_i$ that is scheduled picks a value $x_i$ that is less than or equal to $M_t - S$, draws that much more of the resource, and updates the global variable $S$. As before, this is also how a process reduces the amount of resource it draws.
  - If the value of the global variable $M_t$ changes while this loop is being executed, control exits from the loop.
B. If $M_t$ is less than $S$, i.e., if more resource is being drawn than the maximum resource limit, then the bottom **for** loop, which schedules processes in increasing order of priority, is executed.
  - Each process $p_i$ that is scheduled picks a value $y_i$ that is greater than or equal to 0 and less than or equal to $S - M_t$, and reduces its intake by this amount.

- If the value of the global variable $M_t$ changes while this loop is being executed, control exits from the loop.

As with Algorithm 3.1, it is easy to show that this algorithm also satisfies the Maintenance property, and that it also guarantees Weak Recovery if $y_i$ is always greater than 0 and not otherwise. An obvious analogue of Theorem 1 may also be stated for Strong Recovery. However, the following is more significant.

**Theorem 2.** *It is necessary to halt execution of the first* **for** *loop, if running, in case $M_t$ changes, in order to stay true to Proposition 1.*

*Proof.* We can distinguish the following cases, which we consider in sequence.

$M_t$ increases: in this case, more resource is allowed to be drawn than when the loop started, and by Proposition 1, the greatest benefit of this should go to the highest-priority processes. For this reason, it is necessary to quit the loop execution and start it again.

$M_t$ decreases, but is still greater than $S$: in this case, less of the resource is allowed to be drawn than when the loop started, and the highest-priority processes should have the greatest access to this reduced amount. Therefore, it is necessary to quit the loop and start it again.

$M_t$ decreases, and is now less than $S$: in this case, the loop must be quit so that the second loop can be started.                                                      □

We can also state the obvious analogue of this. The proof is omitted.

**Theorem 3.** *It is necessary to halt execution of the second* **for** *loop, if running, in case $M_t$ changes, in order to stay true to Proposition 1.*

# 4   Conclusions and Future Work

The present work has attempted to put forth a framework to reason about a class of real-life demand-management problems. However, the theory so far is quite rudimentary, and many details need to be addressed. Specifically, work on the other demand management problems besides Maximum Demand is needed. Even in the context of Maximum Demand, wait-free algorithms and algorithms on specific distributed topologies, noting real-life constraints, and technologies such as Smart Grids [15], need to be designed. Design of processes that are well-behaved in the sense described in Section 3.3 is a very non-trivial but important task, one that will doubtless involve significant understanding of related fields like power engineering.

Some pertinent issues that have not been considered in this paper at all include: multiple types of consumable resources (rather than just the one resource such as electrical power); multiple sources with varying limits on each source; and tolerance of process failures.

# References

1. Alyfantis, G., Hadjiefthymiades, S., Merakos, L.: An Overlay Smart Spaces System for Load Balancing in Wireless LANs. Mobile Networks and Applications 11(2), 241–251 (2006)
2. Arthur, W.B.: Inductive reasoning and bounded rationality. American Economic Review 84(2), 406–411 (1994)
3. Cournot, A.A.: Recherches sur les Principes Mathématiques de la Théorie des Richesses. Calmann-Lévy, Paris (1974); Originally published in (1838), this was re-printed (H. Guitton, ed.) as part of a collection titled Les Fondateurs
4. Edmonds, B.: Gossip, sexual recombination and the El Farol bar: modelling the emergence of heterogeneity. Journal of Artificial Societies and Social Simulation 2(3) (October 1999)
5. Enumula, P.K., Rao, S.: The Potluck Problem. Economics Letters 107(1), 10–12 (2010) doi:10.1016/j.econlet.2009.12.011
6. Goh, C.H.K., Apt, J.: Consumer strategies for controlling electric water heaters under dynamic pricing. Carnegie Mellon Electricity Industry Center Working Paper, CEIC-04-02 (2004), http://www.cmu.edu/electricity
7. Grebla, H., Cenan, C.: Distributed database replication—a game theory? In: Seventh International Symposium on Symbolic and Numeric Algorithms for Scientific Computing (SYNASC 2005), September 2005, pp. 25–29 (2005)
8. Greenwald, A.R., Farago, J., Hall, K.: Fair and Efficient Solutions to the Santa Fe Bar Problem. In: Grace Hopper Celebration of Women in Computing (2002)
9. Greenwald, A.R., Mishra, B., Parikh, R.: The Santa Fe Bar Problem revisited: Theoretical and practical implications (April 1998) (unpublished manuscript)
10. Hardin, G.J.: The tragedy of the commons. Science 162, 1243–1248 (1968)
11. Hill, T.T.: Managing scarcity: Changing the paradigm for sustainable resource management. In: Proceedings of the Water Environment Federation, pp. 2659–2665 (2007)
12. Hines, P., Laio, H., Jia, D., Talukdar, S.: Autonomous agents and cooperation for the control of cascading failures in electric grids. In: 2005 IEEE International Conference on Networking, Sensing and Control, March 2005, pp. 273–278 (2005)
13. Klein, R., Day, P., Redmayne, S.: Managing Scarcity: Priority Setting and Rationing in the National Health Service (State of Health). Open University Press (1996)
14. Littlestone, N., Warmuth, M.K.: The weighted majority algorithm. Information and Computation 108(2), 212–261 (1994)
15. Mazza, P.: Powering Up the Smart Grid: A Northwest Initiative for Job Creation, Energy Security and Clean, Affordable Electricity. Climate Solutions (2005), http://www.climatesolutions.org
16. Ranganathan, P.: Recipe for efficiency: Principles of power-aware computing. Commun. of the ACM 53(4), 60–67 (2010)
17. Singh, N., Rao, S.: Modeling and reducing power consumption in large IT systems. In: 4th Annual IEEE International Systems Conference (IEEE SysCon 2010), San Diego, CA (April 2010)
18. Spees, K., Lave, L.: Demand response and electricity market efficiency. Carnegie Mellon Electricity Industry Center Working Paper CEIC-07-01 (January 2007), http://www.cmu.edu/electricity

# Modified Bias Field Fuzzy C-Means for Effective Segmentation of Brain MRI

S.R. Kannan[1,2], S. Ramathilagam[3], and R. Pandiyarajan[2]

[1] Department of Electrical Engineering, NCKU, Tainan 70701, Taiwan
[2] Department of Mathematics, Gandhigram Rural University, Gandhigram 624 302, India
[3] Department of Engineering Science, NCKU, Tainan 70701, Taiwan
s.r.kannan@ruraluniv.ac.in

**Abstract.** In recent day, segmentation of brain Magnetic resonance Image (MRI) with bias field correction is challenging and unavoidable in high magnetic imaging. The brain MRI is affected by bias field that causes the undesired effect of quantitative image analysis. The removal of bias field distortion is useful in segmenting medical images for proper study of medical images. In this paper, we propose three new Fuzzy c-Means (FCM) algorithms namely Robust Gaussian based Weighted Bias Field FCM [RGBFCM], Spatial constraint Gaussian based bias field FCM [GBFCM_S], Novel Penalized Gaussian based Bias field FCM [NPGBFCM] in order to remove bias field distortion and to obtain well segmentation result. The proposed methods are capable to deal with the intensity in-homogeneities and noisy image effectively. Further, to reduce the number of iterations, the proposed algorithms initialize the centroid using *dist-max* initialization algorithm before the execution of algorithms iteratively. To show the performance of proposed methods, this paper applies them to segmentation of brain MRIs and compares the results of our proposed methods with other reported methods. The segmentation accuracy of proposed method is validated by using Silhouette method. The experimental results on real T1-T2 weighted and simulated brain MRIs show that our methods are superior in providing better segmentation results than standard fuzzy c-means based algorithms.

**Keywords:** Brain MRI, FCM, Segmentation, Data analysis, Modified FCM, Clustering.

## 1 Introduction

Image segmentation is one of the most intricate and challenging problems in the field of image processing. Generally, it is used in a different field like robot vision, object recognition, geographical imaging and medical imaging. Especially, the applications of medical image segmentation such as surgical planning, navigation, simulation, diagnosis, and therapy evaluation are all receiving the benefits from segmentation of anatomical structures in medical images. Typically, the image segmentation [10, 23] is defined as the process of assigning a label to every pixel in an image such that pixels with the same label share certain visual characteristics. For the reason that of the

M.L. Gavrilova et al. (Eds.): Trans. on Comput. Sci. VIII, LNCS 6260, pp. 127–145, 2010.

advantages of magnetic resonance imaging  in enhancing the contrast between the tissues over other diagnostic imaging [27], the majority of researches in medical imaging segmentation concerns the MR images for diagnosing cancer related diseases.

The effectiveness of MRI segmentation [20] is declined by spatial intensity non-uniformity [12, 22], also called as bias field. The bias field (intensity in-homogeneity) is induced by the low radio- frequency coil in MRI [6] that reduces the qualitative and quantitative analyses of MR images and it is difficult in most of the medical proceedings for diagnosis, such as the tissue-segmentation, affecting the quantitative image analysis.

So many researchers in the field of medical image segmentation have paid much attention on unsupervised segmentation technique, especially on Fuzzy c-Means algorithm. The Fuzzy C-Means algorithm is an extension of hard c-mans algorithm. It assigns each pixel to multiple clusters with different meaningful membership grade which lies between 0 and 1, while hard c-means assigns each pixel to exactly only one cluster with membership grade 0 or 1. It makes the fuzzy segmentation methods be able to retain more information from the original image than the hard segmentation methods.

However, non impractical conditions lead to the presence of intensity inhomogeneities which terminate in intensity overlapping of different tissues. Further, in conventional FCM approach, each pixel is dealt as a separate unit, independent of its spatial information. This leads to noisy segmented results.

To solve these problems, many researchers developed new modified fuzzy c-means which is incorporated by the concepts Kernel trick, bias field correction, additional term, penalty term, entropy term and spatial neighbourhood information terms, etc. To improve the robustness and segmentation ability, Zhang and Chen [30] proposed kernel based FCM with spatial constraint term (KFCM_S) which is based on the concept of kernel trick and neighbourhoods of pixels. In [25], Siyal and Lin Yu introduced modified FCM for automated segmentation of medical images to deal the intensity inhomogeneities and Gaussian noise effectively. The modified FCM which incorporates the spatial information into the membership function for clustering was proposed by Chuang et al [8] to reduce the spurious blobs and removes the noisy spots in the images. Hou et al [11] presented a regularized fuzzy c-means clustering method for brain tissue segmentation from magnetic resonance images in which regularizer of the total variation type is explored. For the implementation of a speedup scheme for kernel clustering and an approach for correcting spurious intensity variation of MRI images, a fast spatially constrained kernel clustering algorithm was proposed for segmenting MRIs in [21]. In order to overcome the drawbacks of computationally time taking and lacks enough robustness to noise and outliers, Yang and Tsai [28] introduced Gaussian kernel-based fuzzy c-means algorithm (GKFCM) with a spatial bias correction. Zanaty et al [29] proposed alternative Kernelized FCM algorithms (KFCM) that could improve MRI segmentation. This algorithm incorporates the spatial information into the membership function of FCM. An adaptive weighted averaging FCM (AWA-FCM) in which the spatial influence of the neighbouring pixels on the central pixel is included, was developed by Jiayin Kang et al [13]. To enhances the smoothness towards piecewise-homogeneous segmentation and reduces the edge blurring effect, Wang et al [26] proposed adaptive spatial information-theoretic clustering (ASIC) algorithm which is obtained by incorporating spatial constrains to

FCM. To remove the bias field, Karan Sikka et al [19] developed new modified FCM algorithm for automated segmentation of medical images. Shen et al. [24] presented a method is called improved fuzzy segmentation (IFS) algorithm to correct the intensity non-uniformity during segmentation.

Although the above algorithms produce good results in image segmentation, they have some shortcomings such as Random initialization of centroid of clusters, over sensitivity to noise, intensity in homogeneity and large amount of storage space which is also a flaw of many other intensity based segmentation methods.

To overcome these draw backs, this paper propose some improved fuzzy c –means clustering techniques for intensity in-homogeneities estimation and segmentation of brain MRIs. Further, this paper incorporates the novel spatial constraint term, penalty term and neighbourhood attraction into the effective objective function of proposed methods to smooth the boundaries between two tissue classes of brain MRIs and also to reduce the effect of intensity inhomogeneity. Since the proposed methods reduce heavy noise and remove the bias field, it enhances the robustness of the original clustering algorithms to noise and outliers, and still retains the computational simplicity. Moreover, in order to reduce the number of iteration and avoid the effect of outliers in initial centroid of clusters, the algorithms initialize the initial centroid of clusters using *dist-max* initialization method.

The paper is organized as follows. Section 2 reviews the standard FCM and the concept of bias field. Section 3 derives the proposed algorithms of this paper and presents the *dist-max* initialization method.  In section 4, the segmentation results are discussed and compared with other reported techniques. Finally conclusions of this paper are summarized in Section 5.

## 2   FCM and Our Proposed Method

### 2.1   Fuzzy C-Means

Fuzzy C-means (FCM) clustering algorithm developed in the 1970s (Dunn [9]) and extended later (Bezdek [2, 3, 4], Bezdek *et al* [5]). For partitioning the dataset $X=\{x_1,x_2,...., x_n\}$ into $c$ subsets with different meaningful membership grade $u_{ik}$, fuzzy c-means algorithm is minimized the following objective function. The number of clusters $c$ is normally passed as an input parameter.

$$J_{fcm}(U,V) = \sum_{i=1}^{c}\sum_{k=1}^{n} u_{ik}^{m} \left\| x_k - v_i \right\|^2 \tag{1}$$

where $m$ is fuzzifier value which controls the fuzzy partitions (1 for hard clustering, and increasing for fuzzy clustering ); $u_{ik}$ is the fuzzy membership value of object $k$ in cluster $i$ ; $d_{ik}^2 = \left\| x_k - v_i \right\|^2$ is the Euclidean distance; $v_i$ is centroid of each cluster. $U$ is the fuzzy partition matrix and $V$ is the matrix of centroids of clusters. The above objective function is minimized iteratively to obtain $U$ and $V$ and finally the FCM algorithm is obtained to execute the partition of given dataset. Further, the membership grade of each data satisfies the following conditions

$$0 \le u_{ik} \le 1, \text{ for } 1 \le k \le n, 1 \le i \le c$$

$$0 < \sum_{k=1}^{n} u_{ik} < n \text{ for } 1 \le k \le c$$

$$\sum_{i=1}^{c} u_{ik} = 1 \text{ for } 1 \le k \le n$$

## 2.2  Background

The observed MRI signal is modelled as a product of the true signal generated by the underlying anatomy, and a spatially varying factor called the gain field

$$Y_k = X_k N_k, \ \forall k \in \{1, 2, ..., n\} \tag{2}$$

Where $X_k$ and $Y_k$ are the true and observed intensities at the $k^{th}$ voxel, respectively and $N_k$ is the gain field.

The application of a logarithmic transformation to the intensities allows the artifact to be modelled as an additive bias field [1].

$$y_k = x_k + \omega_k \beta_k, \ \forall k \in \{1, 2, ..., n\} \tag{3}$$

where $x_k$ and $y_k$ are the true and observed log-transformed intensities at the $k^{th}$ voxel, respectively, $\omega_k$ is the weight at the $k^{th}$ voxel and $\beta_k$ is the bias field at the $k^{th}$ voxel. If the gain field is known, then it is relatively easy to estimate the tissue class by applying a conventional intensity-based segmented to the corrected data.

## 2.3  Robust Gaussian Based Weighted Bias Field FCM [RGBFCM]

A new effective objective function is introduced in this subsection for attaining robust segmentation of brain MRIs which are affected by intensity inhomogeneity and heavy noise. In addition, an iterative algorithm for minimizing the objective function is obtained here. The proposed objective function contains the weighted bias field estimator and neighbourhood attractions of pixels to remove the bias field and to reduce the noise levels in images. The Gaussian based distance measure is used to measure the similarity between the pixel $x_k$ and centroid of $i^{th}$ cluster $v_i$ in this method.

The new objective function is

$$J_{rgbfcm}(U, V, \beta) = \sum_{i=1}^{c} \sum_{k=1}^{n} u_{ik}^{m} D(x_k - w_k \beta_k, v_i) + \frac{c}{2} \left[ 1 + \sum_{i=1}^{c} \sum_{k=1}^{n} u_{ik}^{m} \sum_{r \in N_k} D(x_r, v_i) \right] \tag{4}$$

Here $N_k$ represents the set of all neighbouring pixels of given pixel $k$. The weighted bias field estimator regularizes the bias field correction and neighbourhood attraction reduces the heavy noise and other artifacts in the medical images. To perform iterative fuzzy partition of given pixel dataset, the objective function $J_{rgbfcm}$ is minimized in a fashion similar to the standard FCM algorithm with respect to $u_{ik}$, $v_i$ and $\beta_k$ by using Lagrangian multipliers method.

The distance measure is defined as $D(x_k, v_i) = 1 - \exp\left(-\dfrac{\|x_k - v_i\|^2}{\sigma^2}\right)$ where $\sigma$ is adjustable parameter. The same distance measure is used for the other proposed methods.

## A. Bias field estimation

The bias field estimator is used to find the bias field in the corrupted brain MRIs. For obtaining well segmentation result, the bias field is removed from the images using weighted bias field estimation. In this subsection the bias field estimator is derived by minimizing eq. (4) with respect to $\beta_k$.

Taking the first order partial derivative of $J_{rgbfcm}$ with respect to $\beta_k$ and setting the result to zero we have

$$
\frac{\partial J}{\partial \beta_k} = \left[ \sum_{i=1}^{c} \frac{\partial}{\partial \beta_k} \sum_{k=1}^{n} u_{ik}^m \left( 1 - \exp\left( \frac{\|x_k - w_k \beta_k - v_i\|}{\sigma^2} \right) \right) \right.
$$
$$
\left. + \frac{c}{2} \left[ 1 + \sum_{i=1}^{c} \sum_{k=1}^{n} u_{ik}^m \sum_{r \in N_k} \left( 1 - \exp\left( \frac{\|x_r - v_i\|}{\sigma^2} \right) \right) \right] \right] = 0
$$

(5)

Since only the $k^{th}$ term in the second summation depends on $\beta_k$ we have:

$$
\left[ \sum_{i=1}^{c} u_{ik}^m \left( -\exp\left( \frac{\|x_k - w_k \beta_k - v_i\|}{\sigma^2} \right) \right) \left( -\frac{2}{\sigma^2} (x_k - w_k \beta_k - v_i) \right) (-1) \right]_{\beta_k = \beta_k^*} = 0 \quad (6)
$$

Differentiating the distance expression, we obtain:

$$
\left[ \begin{array}{l} \displaystyle\sum_{i=1}^{c} u_{ik}^m \left( (1 - D(x_k - w_k \beta_k, v_i))(x_k - v_i) \right) - \\[4mm] \displaystyle\sum_{i=1}^{c} u_{ik}^m \left( (1 - D(x_k - w_k \beta_k, v_i))(w_k \beta_k) \right) \end{array} \right]_{\beta_k = \beta_k^*} = 0 \quad (7)
$$

The zero-gradient condition for the bias- field estimator is expressed as:

$$
\beta_k^* = \frac{1}{w_k} \left[ x_k - \frac{\displaystyle\sum_{i=1}^{c} u_{ik}^m \left( 1 - D(x_k - w_k \beta_k, v_i) \right)(x_k - v_i)}{\displaystyle\sum_{i=1}^{c} u_{ik}^m \left( 1 - D(x_k - w_k \beta_k, v_i) \right)} \right] \quad (8)
$$

$\omega_k \in (0,1)$ is the weight. Suppose taking 10 data, the weight is defined as $\omega_1 = 0.008$ and $\omega_2, ..., \omega_{10}$ is assigned by increasing 0.001.

## B. Centroid of clusters updating

To obtain updating equation for centroid of clusters, the objective function $J_{rgbfcm}$ is differentiated partially with respect to $v_i$ and the zero gradient condition is obtained for it as follows.

$$\left[ \sum_{k=1}^{n} u_{ik}^{m} \left( \begin{array}{c} (1 - D(x_k - w_k\beta_k, v_i)) + \\ \dfrac{c}{2} \sum_{k=1}^{n} u_{ik}^{m} \sum_{r \in N_k} (1 - D(x_r, v_i)) \end{array} \right) (x_k - w_k\beta_k - v_i) \right]_{v_i = v_i^*} = 0 \qquad (9)$$

By simplifying eq. (9), we get the equation for $v_i$ as

$$v_i^* = \frac{\sum_{k=1}^{n} u_{ik}^{m} \left[ (1 - D(x_k - w_k\beta_k, v_i))(x_k - w_k\beta_k) + \dfrac{c}{2} \sum_{k=1}^{n} u_{ik}^{m} \sum_{r \in N_k} (1 - D(x_r, v_i)) x_r \right]}{\sum_{k=1}^{n} u_{ik}^{m} \left[ \left[ (1 - D(x_k - w_k\beta_k, v_i)) + \dfrac{c}{2} \sum_{k=1}^{n} u_{ik}^{m} \sum_{r \in N_k} (1 - D(x_r, v_i)) \right] \right]} \qquad (10)$$

This updating centroid of clusters leads the finest final cluster and thus causes the best fuzzy partition.

## C. Evaluation of Membership grades

To obtain the estimator for membership grade of each data point, the objective function $J_{rgbfcm}$ is partially differentiated with respect to $u_{ik}$ subject to the constraints of membership grade. This work is carried out through the Lagrangian multiplier's method.

The Lagrangian function of eq. (4) is

$$L_{rgbfcm}(U, V, \beta, \lambda) = \sum_{i=1}^{c} \sum_{k=1}^{n} u_{ik}^{m} D(x_k - w_k\beta_k, v_i) +$$
$$\frac{c}{2} \left[ 1 + \sum_{i=1}^{c} \sum_{k=1}^{n} u_{ik}^{m} \sum_{r \in N_k} D(x_r, v_i) \right] - \sum_{k=1}^{n} \lambda_k \left( \sum_{i=1}^{c} u_{ik} - 1 \right) \qquad (11)$$

Taking the first order partial derivative of $L_{gbfcm}$ with respect to $u_{ik}$ and setting the result to zero, we have

$$\frac{\partial L_{rgbfcm}}{\partial u_{ik}} = \left[ mu_{ik}^{m-1}D(x_k - w_k\beta_k, v_i) + \frac{c}{2}mu_{ik}^{m-1}\sum_{r\in N_k}D(x_r, v_i) - \lambda_k \right]_{u_{ik}=u_{ik}^*} = 0 \quad (12)$$

$$u_{ik}^{m-1}\left[ D(x_k - w_k\beta_k, v_i) + \frac{c}{2}\sum_{r\in N_k}D(x_r, v_i) \right]_{u_{ik}=u_{ik}^*} = \frac{\lambda_k}{m} \quad (13)$$

Solving for $u_{ik}$ we get

$$u_{ik} = \left(\frac{\lambda_k}{m}\right)^{1/m-1} \left( \frac{1}{\left[ D(x_k - w_k\beta_k, v_i) + \frac{c}{2}\sum_{r\in N_k}D(x_r, v_i) \right]} \right)^{1/m-1} \quad (14)$$

Since $\sum_{j=1}^{c} u_{jk} = 1, \forall k$ we have

$$\left(\frac{\lambda_k}{m}\right)^{1/m-1} = \frac{1}{\sum_{j=1}^{c}\left( \frac{1}{D(x_k - w_k\beta_k, v_j) + \frac{c}{2}\sum_{r\in N_k}D(x_r, v_j)} \right)^{1/m-1}} \quad (15)$$

Substituting the above equation into eq. (14), the zero-gradient condition for the membership grade estimator can be rewritten as:

$$u_{ik} = \frac{1}{\sum_{j=1}^{c}\left( \frac{D(x_k - w_k\beta_k, v_i) + \frac{c}{2}\sum_{r\in N_k}D(x_r, v_i)}{D(x_k - w_k\beta_k, v_j) + \frac{c}{2}\sum_{r\in N_k}D(x_r, v_j)} \right)^{1/m-1}} \quad (16)$$

The effective membership grade provides the well fuzzy partition and it avoids the overlapping of clustering in final result.

The above minimization problem can be summarized as algorithm for estimating the weighted bias field and segmenting brain MRIs into different tissue classes

*Step 1:* Assign initial centroid of clusters using dist-max initialization algorithm. Fix the value for $c$, $\{\omega_k\}_{k=1}^{n}$ and set the initial value for $\{\beta_k\}_{k=1}^{n}$ which are equal to very small values.

*Step 2:* Compute the membership grade using eq. (16)

*Step 3:* Update the centroid of the clusters using eq. (10)

*Step 4:* Update the bias field estimator using eq. (8)

*Step 5:* Repeat steps 2–4 until termination. The termination criterion is as follows

$$\left\| u_{ik}^{(t+1)} - u_{ik}^{(t)} \right\| < \varepsilon$$

$\varepsilon$ is a small number that can be set during the initialization process.

## 2.4 Spatial Constraint Gaussian Based Bias Field FCM [GBFCM_S]

For automatic effective segmentation of brain MRIs and subsequent removal of the bias field, this section proposes a novel modified objective function which is incorporated by the bias field estimation and spatial constraint term. The Gaussian based distance is used to deal the nonlinear shaped dataset in this method. The influential updating equation for bias field estimator, centroid of clusters and membership grade are obtained by minimization of the new objective function.

The new objective function of bias field FCM as follows

$$J_{gbfcm\_s}(U,V,\beta) = \sum_{i=1}^{c}\sum_{k=1}^{n} u_{ik}^{m} D(x_k - \beta_k, v_i) + (1-\gamma)\left[ \sum_{i=1}^{c}\sum_{k=1}^{n} u_{ik}^{m} D(\overline{x_k} - \beta_k, v_i) \right] \quad (17)$$

The $\overline{x}_k$ is known as the average of spatial neighbouring pixels of $k^{th}$ pixel. In this method, the spatial constraint term is assessed to smooth the boundaries between different tissue classes and also to remove pixel level noise present in the segmented results. The parameter $\gamma$ controls the effect of the noise on each pixel to have desirable membership grade.

## A. Bias field Estimation

In order to deal the inhomogeneous radio-frequency in the brain MRIs, the bias field estimator is derived through the minimization process of eq. (17). The first order partial derivative is taken on the $J_{gbfcm\_s}$ with respect to $\beta_k$ and setting the result to zero. Then we can obtain

$$\frac{\partial J}{\partial \beta_k} = \left[ \begin{array}{c} \sum_{i=1}^{c} \frac{\partial}{\partial \beta_k} \sum_{k=1}^{n} u_{ik}^{m}\left(1-\exp\left(\frac{\|x_k - \beta_k - v_i\|}{\sigma^2}\right)\right) + \\ \frac{c}{2}\left[ \sum_{i=1}^{c}\sum_{k=1}^{n} u_{ik}^{m}\left(1-\exp\left(\frac{\|\overline{x_k} - \beta_k - v_i\|}{\sigma^2}\right)\right)\right] \end{array} \right]_{\beta_k = \beta_k^*} = 0 \quad (18)$$

By solving eq. (18), the updating equation for the bias- field estimator is obtained as

$$\beta_k^* = \frac{\sum\limits_{i=1}^{c} u_{ik}^m \left( \left(1 - D(x_k - \beta_k, v_i)\right)(x_k - v_i) + \left(1 - D(\overline{x}_k - \beta_k, v_i)\right)(\overline{x}_k - v_i) \right)}{\sum\limits_{i=1}^{c} u_{ik}^m \left( \left(1 - D(x_k - \beta_k, v_i)\right) + \left(1 - D(\overline{x}_k - \beta_k, v_i)\right) \right)} \tag{19}$$

The bias field estimator is essentially used to deduct the effect of intensity inhomogeneity in the corrupted MRIs.

## B. Prototype updating

For obtaining well segmentation result, we should have the proper final clusters. The proficient updating equation of centroid of clusters is acquired by minimizing the objective function $J_{gbfcm\_s}$ with respect to the centroid of clusters $v_i$.

Taking the partial derivative of $J_{gbfcm\_s}$ with respect to $v_i$ and setting the result to zero we have

$$\left[ \sum_{k=1}^{n} u_{ik}^m \begin{pmatrix} \left(1 - D(x_k - \beta_k, v_i)\right)(x_k - \beta_k - v_i) + \\ (1-\gamma)\left(1 - D(\overline{x}_k - \beta_k, v_i)\right)(\overline{x}_k - \beta_k - v_i) \end{pmatrix} \right]_{v_i = v_i^*} = 0 \tag{20}$$

Solving for $v_i$, we can get

$$v_i^* = \frac{\sum\limits_{k=1}^{n} u_{ik}^m \left( \left(1 - D(x_k - \beta_k, v_i)\right)(x_k - \beta_k) + (1-\gamma)\left(1 - D(\overline{x}_k - \beta_k, v_i)\right)(\overline{x}_k - \beta_k) \right)}{\sum\limits_{k=1}^{n} u_{ik}^m \left( \left(1 - D(x_k - \beta_k, v_i)\right) + (1-\gamma)\left(1 - D(\overline{x}_k - \beta_k, v_i)\right) \right)} \tag{21}$$

This updating equation for centroid of clusters gives the effective centroids and thus the algorithm obtains the final clusters with less number of iterations.

## C. Membership evaluation

The objective function $J_{gbfcm\_s}$ is minimized by using the Lagrangian multiplier's method subject to the constraint of membership grade for getting the dexterous estimator of membership grade.

The Lagrangian function of eq. (17) is given by

$$L_{gbfcm\_s}(U,V,\beta,\lambda) = \sum_{i=1}^{c}\sum_{k=1}^{n} u_{ik}^m D(x_k - \beta_k, v_i) + \\ (1-\gamma)\left[ \sum_{i=1}^{c}\sum_{k=1}^{n} u_{ik}^m D(\overline{x}_k - \beta_k, v_i) \right] - \sum_{k=1}^{n} \lambda_k \left( \sum_{i=1}^{c} u_{ik} - 1 \right) \tag{22}$$

To obtain membership grade, the $L_{gbfcm\_s}$ is partially differentiated with respect to $u_{ik}$ and the obtained derivation equate to zero as follows

$$\frac{\partial L_{gbfcm\_s}}{\partial u_{ik}} = \left[ mu_{ik}^{m-1} D(x_k - \beta_k, v_i) + (1-\gamma) mu_{ik}^{m-1} D(\overline{x_k} - \beta_k, v_i) - \lambda_k \right]_{u_{ik} = u_{ik}^*} = 0 \quad (23)$$

$$u_{ik}^{m-1} \left[ D(x_k - \beta_k, v_i) + (1-\gamma) D(\overline{x_k} - \beta_k, v_i) \right]_{u_{ik} = u_{ik}^*} = \frac{\lambda_k}{m} \quad (24)$$

Since $\sum_{j=1}^{c} u_{jk} = 1, \forall k$ we have

$$\left( \frac{\lambda_k}{m} \right)^{1/m-1} = \frac{1}{\sum_{j=1}^{c} \left( \frac{1}{D(x_k - \beta_k, v_j) + (1-\gamma) D(\overline{x_k} - \beta_k, v_j)} \right)^{1/m-1}} \quad (25)$$

Substituting the above equation into eq. (24), the zero-gradient condition for the membership estimator can be rewritten as

$$u_{ik} = \frac{1}{\sum_{j=1}^{c} \left( \frac{D(x_k - \beta_k, v_i) + (1-\gamma) D(\overline{x_k} - \beta_k, v_i)}{D(x_k - \beta_k, v_j) + (1-\gamma) D(\overline{x_k} - \beta_k, v_j)} \right)^{1/m-1}} \quad (26)$$

This membership grade presents the effective fuzzy partition of pixel dataset and the borders between the tissues are very clear. This cause well segmentation result of MRIs which are severely affected by intensity inhomogeneity and heavy noise.

The above iterative minimization process for estimating the bias field and segmenting brain MRIs into different tissue classes can be carried out by the following steps:

Step 1: Select initial centroid of clusters by using dist-max initialization algorithm and $\gamma \in [0,1]$

Step 2: Calculate the membership using eq. (26).

Step 3: Update the centroid of the clusters using eq. (21)

Step 4: Update the bias field estimation using eq. (19)

Step 5: Repeat steps 2–4 until termination. The termination criterion is as follows

$$\left\| u_{ik}^{(t+1)} - u_{ik}^{(t)} \right\| < \varepsilon$$

The small value for $\varepsilon$ is given during the initialization process.

## 2.5 Novel Penalized Gaussian Based Bias field FCM [NPGBFCM]

To tackle the intensity inhomogeneity and to ensure effective fuzzification, the novel objective function is constructed in this subsection. A membership grade of each pixel not only depends upon its own intensity but also on that of the neighbouring pixels. This method incorporates the novel penalty term that considering the membership grade of neighbouring pixels. The objective function of proposed method is minimized based on the membership grade, centroid of clusters and bias field estimator, and hence the equation for updating membership grade, centroid of clusters and bias field estimator are obtained in the subsequent sections.

The novel objective function of bias field FCM as follows

$$J_{npgbfcm}(U,V,\beta) = \sum_{i=1}^{c}\sum_{k=1}^{n} u_{ik}^m D(x_k - \beta_k, v_i) + \sum_{i=1}^{c}\sum_{k=1}^{n} u_{ik}^m \left( \sum_{r \in N_k} \sum_{s \in M_i} (1 - u_{rs})^m \right) \quad (27)$$

Here $N_k$ is the set of neighbourhood of $k^{th}$ pixel and $M_i = \{1, 2, .., c\} - \{i\}$.

The penalty term efficiently remove the intensity inhomogeneity that causes the worst segmentation results of MRIs. The Gaussian based distance function is used to measure the similarity between the pixel and centroid of cluster. This makes the algorithm to ease access on general shaped dataset.

## A. Bias field Estimation

To eliminate the intensity inhomogeneity and noise which are occurred in brain MRIs, the bias field estimator is obtained in this subsection. The objective function eq. (27) is minimized based on $\beta_k$ for getting estimator of bias field.

Taking the derivative of $J_{npgbfcm}$ with respect to $\beta_k$ and setting the result to zero we have

$$\frac{\partial J_{npgbfcm}}{\partial \beta_k} = \left[ \begin{array}{c} \sum_{i=1}^{c} \frac{\partial}{\partial \beta_k} \sum_{k=1}^{n} u_{ik}^m D(x_k - \beta_k, v_i) + \\ \sum_{i=1}^{c}\sum_{k=1}^{n} u_{ik}^m \left( \sum_{r \in N_k} \sum_{s \in M_i} (1 - u_{rs})^m \right) \end{array} \right]_{\beta_k = \beta_k^*} = 0 \quad (28)$$

The zero-gradient condition for the bias- field estimator is expressed as:

$$\beta_k^* = \left[ \frac{\sum_{i=1}^{c} u_{ik}^m (1 - D(x_k - \beta_k, v_i))(x_k - v_i)}{\sum_{i=1}^{c} u_{ik}^m (1 - D(x_k - \beta_k, v_i))} \right] \quad (29)$$

The estimation of bias field biases the solution toward the piecewise-homogeneous labelling and precious in segmenting images which are affected by bias field and heavy noise.

## B. Prototype updating

To obtain the best updating equation of centroid of clusters, we minimize the objective function $J_{npgbfcm}$ with respect to $v_i$. First we take first order partial derivative of $J_{npgbfcm}$ and then zeroing the result as follows

$$\left[\sum_{k=1}^{n} u_{ik}^{m}\left(1 - D(x_k - \beta_k, v_i)\right)(x_k - \beta_k - v_i)\right]_{v_i = v_i^*} = 0 \qquad (30)$$

Solving for $v_i$, we obtain the updating estimator for centroid of clusters as

$$v_i^* = \frac{\sum_{k=1}^{n} u_{ik}^{m}\left(1 - D(x_k - \beta_k, v_i)\right)(x_k - \beta_k)}{\sum_{k=1}^{n} u_{ik}^{m}\left(1 - D(x_k - \beta_k, v_i)\right)} \qquad (31)$$

This equation gives the successive centroid of clusters iteratively while segmenting the corrupted brain MRIs.

## C. Membership evaluation

For obtaining finest fuzzy partition of pixel dataset of brain MRIs, the membership grade for each pixel is obtained from proposed objective function $J_{npgbfcm}$. In order to derive the equation for membership grade, The Lagrangian multiplier's method is used to minimize the objective function $J_{npgbfcm}$. The Lagrangian function of $J_{npgbfcm}$ is shown as

$$L_{npgbfcm}(U, V, \beta) = \sum_{i=1}^{c}\sum_{k=1}^{n} u_{ik}^{m} D(x_k - \beta_k, v_i) +$$
$$\sum_{i=1}^{c}\sum_{k=1}^{n} u_{ik}^{m}\left(\sum_{r \in N_k}\sum_{s \in M_i}(1 - u_{rs})^{m}\right) - \sum_{k=1}^{n}\lambda_k\left(\sum_{i=1}^{c} u_{ik} - 1\right) \qquad (32)$$

Taking the derivative of $L_{npgbfcm}$ with respect to $u_{ik}$ and setting the result to zero, we have

$$\frac{\partial L_{gbfcm\_s}}{\partial u_{ik}} = \left[mu_{ik}^{m-1}D(x_k - \beta_k, v_i) + mu_{ik}^{m-1}\left(\sum_{r \in N_k}\sum_{s \in M_i}(1 - u_{rs})^{m}\right) - \lambda_k\right]_{u_{ik} = u_{ik}^*} = 0 \quad (33)$$

$$u_{ik}^{m-1}\left[D(x_k - \beta_k, v_i) + \left(\sum_{r \in N_k}\sum_{s \in M_i}(1 - u_{rs})^{m}\right)\right]_{u_{ik} = u_{ik}^*} = \frac{\lambda_k}{m} \qquad (34)$$

Since $\sum_{j=1}^{c} u_{jk} = 1, \forall k$ we have

$$\left(\frac{\lambda_k}{m}\right)^{1/m-1} = \frac{1}{\sum_{j=1}^{c}\left[\dfrac{1}{D(x_k - \beta_k, v_j) + \left(\sum_{r \in N_k}\sum_{s \in M_i}(1 - u_{rs})^m\right)}\right]^{1/m-1}} \tag{35}$$

Substituting the above equation into eq. (34), the zero-gradient condition for the membership estimator can be rewritten as

$$u_{ik} = \frac{1}{\sum_{j=1}^{c}\left[\dfrac{D(x_k - \beta_k, v_i) + \left(\sum_{r \in N_k}\sum_{s \in M_i}(1 - u_{rs})^m\right)}{D(x_k - \beta_k, v_j) + \left(\sum_{r \in N_k}\sum_{s \in M_i}(1 - u_{rs})^m\right)}\right]^{1/m-1}} \tag{36}$$

It makes the clear border between tissues while segmenting brain MRIs and proper structure of the clusters.

The process of *NPGBFCM* is carried out through the following steps for estimating the bias field and segmenting real brain MRIs into different tissue classes.

Step1: Choose the initial centroid of clusters using dist-max initialization algorithm and fix the initial value for $\beta_k$

Step2: Evaluate the membership grade using eq. (36)

Step3: Update the centroid of the clusters using eq. (31)

Step4: Update the bias field estimation using eq. (29)

Step5: Repeat steps 2–4 until termination. The termination criterion is as follows

$$\left\| u_{ik}^{(t+1)} - u_{ik}^{(t)} \right\| < \varepsilon$$

During the initialization process of algorithm, the value for $\varepsilon$ is given.

## 3  *dist-max* Initialization Algorithm

Step 1: Let $X = \{x_1, x_2, \ldots, x_n\} \subset R^p$ be a data set, where $p$-Dimension.

Find $s = \left\lfloor \dfrac{n}{c} \right\rfloor$ and $m_1, m_2, \ldots, m_n$, where $m_i = \dfrac{x_{i1} + x_{i2} + \ldots + x_{ip}}{p}$,

$i = 1, 2, ..., n$, $c$ be the number of cluster. Arrange $m_i$'s in ascending order.

Step 2: Rearrange the data matrix in respect of its relabeling mean value. (i.e) $X' = [x'_1, x'_2, ......x'_n]$. Partitioning the data into $c$ groups. First group contains first s data of X'. Second group contains second $s$ data of X'

.

.

.

$(c-1)^{th}$ group contains $(c-1)^{th}$ $s$ data of X'. $c^{th}$ group contains remaining all elements.

Step 3: Making a distance tables that show the distance between the elements within each group. (ie) if group $k = [x_1^k, x_2^k, ......x_n^k]$, the distance table is

|        | $x_1^k$  | $x_2^k$  | · · · · · · · · · · · · · | $x_n^k$  |
|--------|----------|----------|--------------------------|----------|
| $x_1^k$ | $d_{11}^k$ | $d_{12}^k$ |                          | $d_{1n}^k$ |
| $x_2^k$ | $d_{21}^k$ | $d_{22}^k$ |                          | $d_{2n}^k$ |
| .      | .        | .        | · · · · · · · · · · · · · · · |          |
| $x_n^k$ | $d_{n1}^k$ | $d_{n2}^k$ | · · · · · · · · · · · · · · · | $d_{nn}^k$ |

Step 4: Select maximum distance from each distance table of groups. If $d_{ij}^k$ is maximum distance of $k^{th}$ group, find the mean value $M_k$ of the elements $x_i$ and $x_j$. $k^{th}$ cluster center $= M_k$. $k=1, 2,...,c$

## 4  Results and Discussion

In this section, we demonstrate some experimental analysis and results. The objective of these experiments is to determine the excellence of our proposed approach. To compare the segmentation performance of our proposed methods RGBFCM, GBFCM_S and NPGBFCM with the segmentation performance of existed algorithms IFS [23] and KFCM [27], the experimental work is performed on real T1-T2 weighted and simulated brain MRIs shown in Figs. 1(a-b) & Figs. 2(a-b). The additive Gaussian white noise is added with brain T1-T2 weighted and simulated brain medical images for only the purpose of showing the robustness of the proposed algorithms, but in nature the MRI images are not affected by Gaussian noise.

We implement the IFS, KFCM and proposed methods on real T1 and T2 and simulated brain MRIs which are corrupted by Gaussian noise heavily. Figs. 1-2 (c-l) show the segmentation results of IFS, KFCM, and Proposed methods. As shown in Figs. 1-2(c-f), neither IFS nor KFCM can classify the tissues of images properly, while our proposed methods succeeded well in correcting and classifying the tissues of images as shown in Figs 1-2 (g-l). From the images, we can see that IFS and KFCM achieve poor segmentation result and fail to remove the noises, while the proposed methods achieve well segmentation results and have no problem about the effect of Gaussian noise.

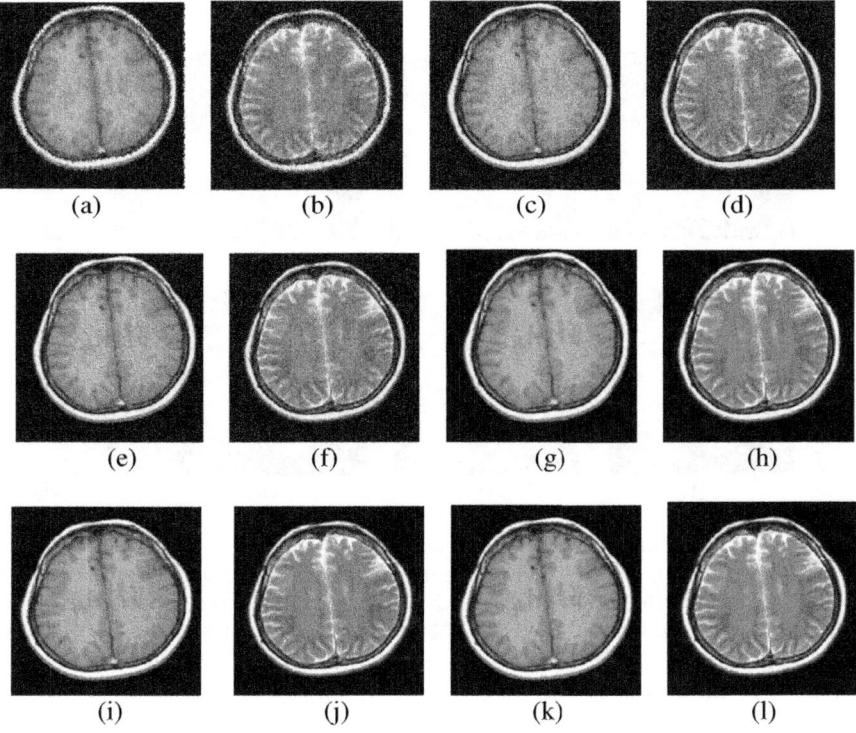

**Fig. 1.** (a) T1 corrupted by Gaussian noise, (b) T2 corrupted by Gaussian noise, (c) T1 Segmented by *IFS* (d) T2 Segmented by *IFS* (e) T1 Segmented by *KFCM* (f) T2 Segmented by *KFCM* (g) T1 Segmented by RGBFCM (h) T2 Segmented by RGBFCM (i) T1 Segmented by GBFCM_S (j) T2 Segmented by GBFCM_S (k) T1 Segmented by NPGBFCM (l) T2 Segmented by NPGBFCM

For quantitative evaluation of performance, the segmentation accuracy is obtained for those five algorithms where segmentation accuracy is defined using silhouette value in [15, 16, 17]. Table 1 shows the segmentation accuracy of the IFS, KFCM and proposed algorithms on two different T1-T2 noisy images. This silhouette average value measures the degree of confidence in the clustering assignment of a particular observation with well clustered observation having values near 1 and poorly clustered observation having value near -1, and it assesses the accuracy of the segmented results.

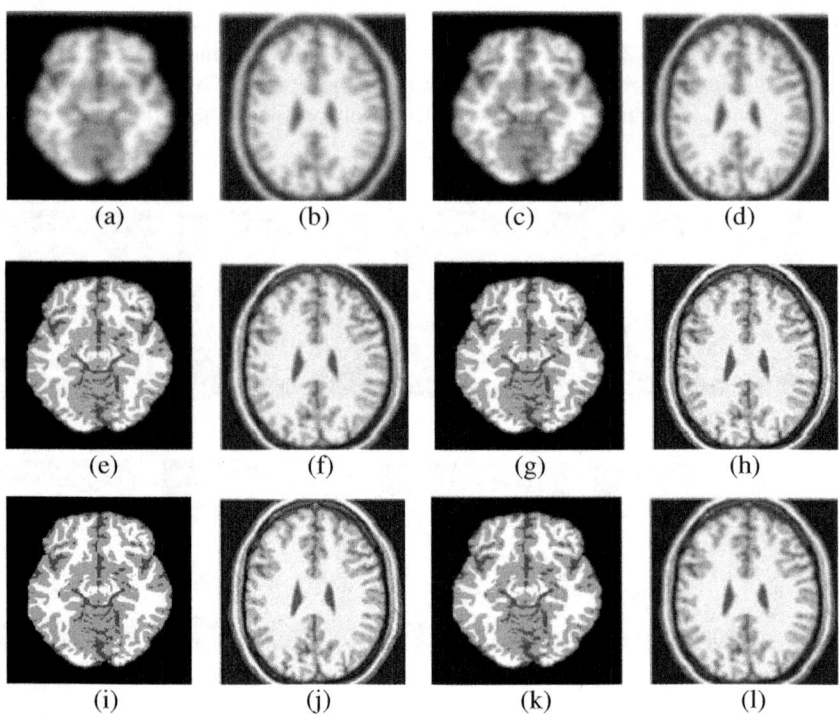

**Fig. 2.** (a) Simulated brain Image-1 corrupted by Gaussian noise (b) Simulated brain Image -2 corrupted by Gaussian noise (c) Image-1 Segmented by *IFS* (d) Image -2 Segmented by *IFS* (e) Image -1 Segmented by *KFCM*, (f) Image -2 Segmented by *KFCM* (g) Image -1 Segmented by RGBFCM (h) Image -2 Segmented by RGBFCM (i) Image -1 Segmented by GBFCM_S (j) Image -2 Segmented by GBFCM_S (k) Image -1 Segmented by NPGBFCM (l) Image -2 Segmented by NPGBFCM

**Table 1.** Segmentation accuracies

| No. of Clusters | Silhouette Value | Accuracy |
|---|---|---|
| IFS | 0.47 | 47% |
| KFCM | 0.52 | 52% |
| Proposed RGBFCM | 0.72 | 72% |
| Proposed GBFCM_S | 0.76 | 76% |
| Proposed NPGBFCM | 0.79 | 79% |

From Table 1, the best clustering validity 79 % was obtained for proposed method NPGBFCM during the experimental work on brain image dataset. It is clear from Figs. 1-2 (g-l) that our proposed method completely succeeded in correcting and classifying the brain data and almost the methods have eliminated completely the effect of noise in images. The other two IFS and KFCM techniques partially corrected the misclassified pixels given in Fig. 1-2 (c-f). From the above experimental results of real T1-T2 weighted and simulated brain medical images, we conclude that the proposed methods performed much better than existed ones. Particularly, NPGBFCM algorithm acquires the best segmentation performance.

## 5   Conclusion

In this paper, we proposed three new Fuzzy C-Means  algorithms namely Robust Gaussian based Weighted Bias Field FCM [RGBFCM], Spatial constraint Gaussian based bias field FCM [GBFCM_S], Novel Penalized Gaussian based Bias field FCM [NPGBFCM] that are capable to deal with the intense in-homogeneities and noisy image effectively. In order to remove bias field distortion and to obtain well segmentation result, the objective function of proposed methods have been  incorporated by the bias field information, neighbourhood attraction, Spatial constraints term, novel penalty term and Gaussian based distance function. The proposed algorithms have initialized the centroid using *dist-max* initialization algorithm before the execution of algorithm iteratively. To show the performance of proposed methods, the algorithms IFS, KFCM and our proposed methods were implemented on real T1-T2 weighted and simulated brain MRIs and the segmentation results of proposed methods compared with results of other existed methods. Further, the Silhouette method was used to evaluate the segmentation accuracies of those five algorithms and our proposed method achieved high segmentation accuracy compared with existed methods. It is clear form the experimental results that our proposed method more robust to the noises and faster than other methods. Particularly, the proposed algorithm NPGBFCM outperformed other methods in many aspects, especially in segmentation accuracy, which is the most important in tissue classification of brain MRIs.

## References

1. Ahmed, M.N., Mohamed, S.M.N., Farag, A.A., Moriarty, T.: A modified fuzzy c-means algorithm for bias field estimation and segmentation of MRI data. IEEE Trans. Med. Imaging 21, 193–199 (2002)
2. Bezdek, J.C.: Fuzzy Mathematics in Pattern Classification. The Institute of Electrical and Electronics Engineers, New York (1973)
3. Bezdek, J.C.: A convergence theorem for the fuzzy data clustering algorithms. IEEE Trans. Pattern Anal. Mach. Intell. PAMI 2, 1–8 (1980)
4. Bezdek, J.C.: Fuzzy Models and Algorithms for Pattern Recognition and Image Processing. Kulwer Academic Publishers, Boston (1999)
5. Bezdek, J.C., Hathaway, R.J., Sabin, M.J., Tucker, W.T.: Convergence theory for fuzzy C-means: counterexamples and repairs. IEEE Trans. Syst. Man Cybern. 17, 873–877 (1987)

6. Cannon, R.L., Dave, J.V., Bezdek, J.C.: Efficient implementation of the fuzzy c- means clustering algorithms. IEEE Transaction on Pattern Analysis and Machine Intelligence PAMI-8(2), 248–255 (1986)
7. Condon, B.R., Patterson, J., Wyper, D.: Image non uniformity in magnetic resonance imaging: Its magnitude and methods for its correction. Br. J. Radiol. 60, 83–87 (1987)
8. Chuang, K.S., et al.: Fuzzy c-means clustering with spatial information for image segmentation. Computerized Medical Imaging and Graphics (30), 9–15 (2006)
9. Dunn, J.C.: A fuzzy relative of the ISODATA process and its use in detecting compact well-separated clusters. J. Cybern. 3, 32–57 (1974)
10. Fu, S.K., Mu, J.K.: A survey on image segmentation. Pattern Recognit. 13, 3–16 (1981)
11. Hou, Z., et al.: Regularized fuzzy c-means method for brain tissue clustering. Pattern Recognition Letters (28), 1788–1794 (2007)
12. Hou, Z.: A Review on MR Image Intensity in homogeneity Correction. International Journal of Biomedical Imaging, 1–11 (2006)
13. Kang, J., et al.: Novel modified fuzzy c-means algorithm with applications. Digital Signal Processing (19), 309–319 (2009)
14. Johnston, B., Atkins, M.S., Mackiewich, B., Anderson, M.: Segmentation of multiple sclerosis lesions in intensity corrected multispectral MRI. IEEE Trans. Med. Imaging 15, 154–169 (1996)
15. Kannan, S.R.: A new segmentation system for brain MR images based on fuzzy techniques. Applied Soft Computing 8(4), 1599–1606 (2008)
16. Kannan, S.R., Sathya, A., Ramathilagam, S.: Effective Fuzzy Clustering Techniques for Segmentation of Breast MRI. International Journal of Soft Computing (Springer Publication). Published online: November 01 (2009) doi: 10.1007/s00500-009-0528-8
17. Kannan, S.R., Ramathilagam, S., Sathya, A., Pandiyarajan, R.: Effective fuzzy c-means based kernel function in segmenting medical images. Journal of Computers in Biology and Medicine (Elsevier) (in Press) doi:10.1016/j.compbiomed.2010.04.001
18. Kapur, T., Grimson, W.E.L., Wells, W.M., Kikinis, R.: Segmentation of brain tissue from magnetic resonance images. Medical Image Analysis 1, 109–127 (1996)
19. Sikka, K., et al.: A fully automated algorithm under modified FCM framework for Improved brain MR image segmentation. Magnetic Resonance Imaging (27), 994–1004 (2009)
20. Leemput, K.V., Maes, F., Vandermeulen, D., Suetens, P.: Automated model based bias field correction of MR images of the brain. IEEE Trans. on Medical Imaging 18, 885–896 (1999)
21. Liao, L., Lin, T., Li, B.: MRI brain image segmentation and bias field correction based on fast spatially constrained kernel clustering approach. Pattern Recognition Letters (29), 1580–1588 (2008)
22. Pham, D.L., Prince, J.L.: Adaptive fuzzy segmentation of magnetic resonance im-ages. IEEE Trans. Med. Imag. 18, 737–752 (1999)
23. Rajapakse, J.C., Krugge, F.: Segmentation of MR images with intensity in homogeneities. Image Vision Comput. 16(3), 165–180 (1998)
24. Shen, S., Sandhan, W.A., Granat, M.H., Sterr, A.: Intensity Non-uniformity correction of Magnetic Resonance Images using a Fuzzy segmentation Algorithm. In: Proceedings of the 27th Annual Conference on IEEE Engineering in Medicine and Biology, Shanghai, China, pp. 3035–3038 (2005)
25. Siyal, M.Y., Yu, L.: An intelligent modified fuzzy c-means based algorithm for bias estimation and segmentation of brain MRI. Pattern Recognition Letters 26, 2052–2062 (2005)

26. Wang, Z.M., et al.: Adaptive spatial information-theoretic clustering for image segmentation. Pattern Recognition (42), 2029–2044 (2009)
27. Wells, W.M., Grimson, W.E.L., Kikinis, R., Jolesz, F.A.: Adaptive segmentation of MRI data. IEEE Trans. Med. Imag. 15, 429–442 (1996)
28. Yang, M.S., Tsai, H.S.: A Gaussian kernel-based fuzzy c -means algorithm with a spatial bias correction. Pattern Recognition Letters (29), 1713–1725 (2008)
29. Zanaty, E.A., Aljahdali, S., Debnath, N.: A kernelized fuzzy c-means algorithm for automatic magnetic resonance image segmentation. Journal of Computational Methods in Sciences and Engineering (9), S123–S136 (2009)
30. Zhang, D.Q., Chen, S.C.: A novel kernelized fuzzy C-means algorithm with application in medical image segmentation. Artificial Intelligence in Medicine 32, 37–50 (2004)

# Visualization of Monotone Data by Rational Bi-cubic Interpolation

Malik Zawwar Hussain[1], Maria Hussain[2], and Muhammad Sarfraz[3]

[1] Department of Mathematics, University of the Punjab, Lahore-Pakistan
malikzawwar@math.pu.edu.pk
[2] Department of Mathematics, Lahore College for Women University,
Lahore-Pakistan
mariahussain_1@yahoo.com
[3] Department of Information Sciences, Adailiya Campus, Kuwait University, Kuwait
prof.m.sarfraz@gmail.com, sarfraz@cfw.kuniv.edu

**Abstract.** The most general piecewise rational cubic function (GPRC) for monotone curve design has been extended to the rational bi-cubic partially blended function to preserve the shape of 3D monotone data. The rational bi-cubic partially blended function involves eight parameters in its description (four along each coordinate axes). Out of these eight shape parameters, four are constrained to preserve the shape of monotone data. The rest of the four parameters are free parameters and have been left free for the users to refine the shape of surface as desired. The developed method not only preserves the monotonicity of the data, but also assures that the visual display is smooth and pleasant.

**Keywords:** Shape preserving, bi-cubic partially blended rational function, monotone surfaces, visualization, 3D data.

## 1 Introduction

Shape preservation of monotone data in the view of monotone curves and surfaces is a fundamental problem in the area of scientific visualization. There are many physical situations where entities only have a meaning when there values are monotone. For example, the data arising from DAC (digital to analog converter) is always monotone. As input code increases in value analog output also increases. The study of tensile strength of material in engineering gives rise to monotone data. Other examples of monotone data are E.S.R. level in cancer patients and blood uric acid level in patients suffering from gout.

The problem of monotonicity has been discussed by various authors in last few years [1-10]. Beatson and Ziegler [1] interpolated monotone data, given on a rectangular grid, with a $C^1$ monotone quadratic spline. They derived necessary and sufficient conditions on functional and derivative values to visualize monotone data. Carlson and Fritsch [2] extended their result of univariate monotone interpolation to bivariate monotone interpolation, for regular data. The interpolating function was determined by first partial derivatives and first mixed

M.L. Gavrilova et al. (Eds.): Trans. on Comput. Sci. VIII, LNCS 6260, pp. 146–155, 2010.

partial derivatives (twist) at mesh points. Necessary and sufficient conditions on these derivatives were derived such that the resulting bi-cubic polynomial is monotone on a single rectangular element. Casciola and Romani [3] presented shape preserving techniques for bivariate NURBS that allowed the user to interactively modify the resulting surface by a set of tension parameters. Clemens and Jütter [4] used *B-H*-curves for modelling ferromagnetic materials in connection with electromagnetic field computations. They presented an approximation technique based on the spline functions and a data dependent smoothing functional. The technique preserved monotonicity as well as smoothness. Costantini [5] proposed solution of problem of existence of monotone and/or convex splines, having degree $n$ and order of continuity $k$, which interpolate to a set of data at knots. The interpolating splines were obtained by using Bernstein polynomials of suitable continuous piecewise linear functions. The presented work are useful in developing algorithms for the construction of shape preserving splines interpolation for arbitrary set of data points. Floater and Peña [6] defined, characterized and compared systems of monotonicity preserving bivariate functions on triangles. Han and Schumaker [7] derived sufficient conditions on the Bézier net of a Bernstein-Bézier polynomial defined on a triangle in the plane to insure that the corresponding surface is monotone. Hussain and Maria [8] used piecewise rational cubic function to visualize monotone data in the view of monotone curves by making constraints on free parameters in the description of rational cubic function. The rational cubic function is extended to rational bi-cubic partially blended function (Coons Patches). Simple constraints were derived on free parameters in the description of rational bi-cubic partially blended patches to visualize the monotone data in the view of monotone surfaces. Sarfraz, Butt and Hussain [10] used a bi-cubic function to visualize the shape of monotone data. In [10] simple constraints were derived on the shape parameters in the description of rational bi-cubic function to preserve the shape of the data.

In this paper, the problem of monotonicity has been addressed using the most general piecewise rational cubic (GPRC) [9]. GPRC provides a variety of shape control parameters that can be sufficient and highly useful for any kind of scientific or mathematical functions. In the case of monotone curve interpolation, we have the simple sufficient data dependent conditions on two free parameters to preserve the shape of monotone data and rest of the two free parameters are used for further refinement of the curves, if needed. This idea has been extended here to a complex problem for the 3D monotone data visualization. The 4-parameter curve description has been generalized to 8-parameter surface description. That is, in the case of monotone data or function, monotone surface interpolation has been achieved with mathematical constraints. We have derived simple sufficient data dependent conditions on four free parameters to preserve the shape of monotone data and rest of four free parameters are used to the further refinement of the surfaces, if needed.

The remainder of the paper is organized as follows. Section 2 describes the most general piecewise rational cubic function (GPRC) [9], used in this paper. In Section 3, the GPRC is extended to the rational bi-cubic partially blended function. In Section 4, simple constraints are derived on free parameters in the description of rational bi-cubic partially blended function to preserve the shape of 3D monotone data. In Section 5, the monotonicity preserving scheme developed in Section 4, is demonstrated with some monotone data. Finally, Section 6 concludes the paper.

## 2   Review of Rational Cubic Function

Let $\{(x_i, f_i), \ i = 0, 1, 2, ..., n\}$ be the given set of data points where $x_0 < x_1 < x_2 < ... < x_n$. The piecewise rational cubic function [9], in its most general form, is defined over each subinterval $I_i = [x_i, x_{i+1}]$ as:

$$S_i(x) = \frac{p_i(\theta)}{q_i(\theta)}, \tag{1}$$

where

$$p_i(\theta) = r_i f_i(1 - \theta)^3 + (u_i f_i + h_i r_i d_i)\theta(1 - \theta)^2 + (v_i f_{i+1} - h_i \omega_i d_{i+1})\theta^2(1 - \theta)$$
$$+ \omega_i f_{i+1}\theta^3,$$
$$q_i(\theta) = r_i(1 - \theta)^3 + u_i\theta(1 - \theta)^2 + v_i\theta^2(1 - \theta) + \omega_i\theta^3,$$
$$\theta = \frac{x - x_i}{h_i}, \ h_i = x_{i+1} - x_i.$$

The rational cubic function (1) has the following properties:

$$S(x_i) = f_i, \ S(x_{i+1}) = f_{i+1}, \ S^{(1)}(x_i) = d_i, \ S^{(1)}(x_{i+1}) = d_{i+1}.$$

Here $S^{(1)}(x)$ denotes the derivative with respect to $x$ and $d_i$ denotes derivative values(given or estimated by some method) at knot $x_i$. $S(x) \in C^1[x_0, x_n]$ freedom of $r_i$, $u_i$, $v_i$ and $\omega_i$ as parameters in the interval $I_i = [x_i, x_{i+1}]$. It has been noted that in each interval, when $r_i = \omega_i = 1$ and $u_i = v_i = 3$, the piecewise rational cubic function (1) reduces to standard cubic Hermite. The following result can be seen in Hussain and Sarfraz [9]:

**Theorem 1.** The piecewise rational cubic function defined in (1) preserves the monotoncity if in each subinterval $I_i = [x_i, x_{i+1}]$ the parameters $u_i$ and $v_i$ satisfy the following conditions:

$$u_i > \frac{r_i d_i}{\Delta_i}, \ v_i > Max\left\{\frac{\omega_i d_{i+1}}{\Delta_i}, \frac{3r_i\omega_i - u_i\omega_i d_{i+1}}{u_i\Delta_i + r_i d_i}\right\},$$

$r_i$, $\omega_i$ are positive real numbers.

## 3   Partially Blended Rational Bi-cubic Function

The piecewise rational cubic function (1) is extended to partially blended bi-cubic function $S(x, y)$ over rectangular domain $D = [a, b] \times [c, d]$. Let $\pi : a =$

$x_0 < x_1 < x_2 < \ldots < x_n = b$ be the partition of $[a, b]$ and $\tilde{\pi} : c = y_0 < y_1 < y_2 < \ldots < y_m = d$ be the partition of $[c, d]$. The rational bi-cubic partially blended function is defined over each rectangular patch $I_{i,j} = [x_i, x_{i+1}] \times [y_j, y_{j+1}]$ where $i = 0, 1, 2, \ldots, n - 1;\ j = 0, 1, 2, \ldots, m - 1$ as:

$$S(x, y) = -AFB^T, \tag{2}$$

where

$$F = \begin{pmatrix} 0 & S(x, y_j) & S(x, y_{j+1}) \\ S(x_i, y) & S(x_i, y_j) & S(x_i, y_{j+1}) \\ S(x_{i+1}, y) & S(x_{i+1}, y_j) & S(x_{i+1}, y_{j+1}) \end{pmatrix},$$

$$A = [-1 \quad a_0(\theta) \quad a_1(\theta)], \quad B = [-1 \quad b_0(\phi) \quad b_1(\phi)],$$

with

$$a_0 = (1 - \theta)^2(1 + 2\theta), \ a_1 = \theta^2(3 - 2\theta), \ b_0 = (1 - \phi)^2(1 + 2\phi), \ b_1 = \phi^2(3 - 2\phi),$$

$$\theta = \frac{x - x_i}{h_i}, \quad \phi = \frac{y - y_j}{\hat{h}_j}.$$

The functions $S(x, y_j)$, $S(x, y_{j+1})$, $S(x_i, y)$ and $S(x_{i+1}, y)$ are rational cubic functions (1) defined over the boundary of rectangular patch $I_{i,j} = [x_i, x_{i+1}] \times [y_j, y_{j+1}]$. These are described in (3)-(6) as follows:

$$S(x, y_j) = \frac{\sum_{i=0}^{3}(1 - \theta)^{3-i}\theta^i A_i}{q_1(\theta)}, \tag{3}$$

with

$$A_0 = r_{i,j}F_{i,j}, \ A_1 = u_{i,j}F_{i,j} + r_{i,j}h_i F_{i,j}^x, \ A_2 = v_{i,j}F_{i+1,j} - \omega_{i,j}h_i F_{i+1,j}^x,$$
$$A_3 = \omega_{i,j}F_{i+1,j}, \ q_1(\theta) = r_{i,j}(1 - \theta)^3 + u_{i,j}\theta(1 - \theta)^2 + v_{i,j}\theta^2(1 - \theta) + \omega_{i,j}\theta^3.$$

$$S(x, y_{j+1}) = \frac{\sum_{i=0}^{3}(1 - \theta)^{3-i}\theta^i B_i}{q_2(\theta)}, \tag{4}$$

with

$$B_0 = r_{i,j+1}F_{i,j+1}, \ B_1 = u_{i,j+1}F_{i,j+1} + r_{i,j+1}h_i F_{i,j+1}^x,$$
$$B_2 = v_{i,j+1}F_{i+1,j+1} - \omega_{i,j+1}h_i F_{i+1,j+1}^x, \ B_3 = \omega_{i,j+1}F_{i+1,j+1},$$
$$q_2(\theta) = r_{i,j+1}(1 - \theta)^3 + u_{i,j+1}\theta(1 - \theta)^2 + v_{i,j+1}\theta^2(1 - \theta) + \omega_{i,j+1}\theta^3.$$

$$S(x_i, y) = \frac{\sum_{i=0}^{3}(1 - \phi)^{3-i}\phi^i C_i}{q_3(\phi)}, \tag{5}$$

with

$$C_0 = \hat{r}_{i,j}F_{i,j}, \ C_1 = \hat{u}_{i,j}F_{i,j} + \hat{r}_{i,j}\hat{h}_j F_{i,j}^y, \ C_2 = \hat{v}_{i,j}F_{i,j+1} - \hat{\omega}_{i,j}\hat{h}_j F_{i,j+1}^y,$$
$$C_3 = \hat{\omega}_{i,j}F_{i,j+1}, \ q_3(\phi) = \hat{r}_{i,j}(1 - \phi)^3 + \hat{u}_{i,j}\phi(1 - \phi)^2 + \hat{v}_{i,j}\phi^2(1 - \phi) + \hat{\omega}_{i,j}\phi^3.$$

$$S(x_{i+1}, y) = \frac{\sum_{i=0}^{3}(1 - \phi)^{3-i}\phi^i D_i}{q_4(\phi)}, \tag{6}$$

with

$$D_0 = \hat{r}_{i+1,j}F_{i+1,j}, \; D_1 = \hat{u}_{i+1,j}F_{i+1,j} + \hat{r}_{i+1,j}\hat{h}_j F_{i+1,j}^y,$$

$$D_2 = \hat{v}_{i+1,j}F_{i+1,j+1} - \hat{\omega}_{i+1,j}\hat{h}_j F_{i+1,j+1}^y, \; D_3 = \hat{\omega}_{i+1,j}F_{i+1,j+1},$$

$$q_4(\phi) = \hat{r}_{i+1,j}(1 - \phi)^3 + \hat{u}_{i+1,j}\phi(1 - \phi)^2 + \hat{v}_{i+1,j}\phi^2(1 - \phi) + \hat{\omega}_{i+1,j}\phi^3.$$

## 4   Monotone Surface Interpolation

Let $\{(x_i, y_j, F_{i,j}), \; i = 0, 1, 2, \ldots, n - 1; j = 0, 1, 2, \ldots, m - 1\}$ be the mono-
tone data defined over the rectangular grid $I_{i,j} = [x_i, x_{i+1}] \times [y_j, y_{j+1}], \; i = 0, 1, 2, \ldots, n - 1; \; j = 0, 1, 2, \ldots, m - 1$ such that

$$F_{i,j} < F_{i+1,j}, \; F_{i,j} < F_{i,j+1}, \; F_{i,j}^x > 0, \; F_{i,j}^y > 0, \; \Delta_{i,j} > 0, \; \hat{\Delta}_{i,j} > 0, \; \forall i, j.$$

Then, due to the fact that the bi-cubic partially blended surface patch inher-
its all the properties of network of boundary curves [3], the bi-cubic partially
blended surface patch (2) is monotone if boundary curves $S(x, y_j)$, $S(x, y_{j+1})$,
$S(x_i, y)$ and $S(x_{i+1}, y)$ defined in (3)-(6) are monotone. That is,

$S(x, y_j)$ is monotone if $S^{(1)}(x, y_j) > 0$ i.e.

$$S^{(1)}(x, y_j) = \frac{\sum_{i=0}^{5}(1 - \theta)^{5-i}\theta^i E_i}{(q_1(\theta))^2} > 0, \tag{7}$$

with

$$E_0 = r_{i,j}^2 F_{i,j}^x, \; E_1 = 2r_{i,j}v_{i,j}\Delta_{i,j} - 2r_{i,j}\omega_{i,j}F_{i+1,j}^x + r_{i,j}^2 F_{i,j}^x,$$

$$E_2 = (3r_{i,j}\omega_{i,j} + 2r_{i,j}v_{i,j} + u_{i,j}v_{i,j})\Delta_{i,j} - (u_{i,j}\omega_{i,j} + 2r_{i,j}\omega_{i,j})F_{i+1,j}^x$$
$$\qquad -r_{i,j}v_{i,j}F_{i,j}^x,$$

$$E_3 = (3r_{i,j}\omega_{i,j} + 2u_{i,j}\omega_{i,j} + u_{i,j}v_{i,j})\Delta_{i,j} - (2r_{i,j}\omega_{i,j} + r_{i,j}v_{i,j})F_{i,j}^x$$
$$\qquad -u_{i,j}\omega_{i,j}F_{i+1,j}^x,$$

$$E_4 = 2u_{i,j}\omega_{i,j}\Delta_{i,j} - 2r_{i,j}\omega_{i,j}F_{i,j}^x + \omega_{i,j}^2 F_{i+1,j}^x, \; E_5 = \omega_{i,j}^2 F_{i+1,j}^x.$$

Thus, $S^{(1)}(x, y_j) > 0$ if $\sum_{i=0}^{5}(1 - \theta)^{5-i}\theta^i E_i > 0$. $\sum_{i=0}^{5}(1 - \theta)^{5-i}\theta^i E_i > 0$ if
$E_i > 0, \; i = 0, 1, 2, 3, 4, 5$. $E_i > 0, \; i = 0, 1, 2, 3, 4, 5$ if

$$u_{i,j} > \frac{r_{i,j}F_{i,j}^x}{\Delta_{i,j}}, \; v_{i,j} > Max\left\{\frac{\omega_{i,j}F_{i+1,j}^x}{\Delta_{i,j}}, \frac{3r_{i,j}\omega_{i,j} - u_{i,j}\omega_{i,j}F_{i+1,j}^x}{u_{i,j}\Delta_{i,j} + r_{i,j}F_{i,j}^x}\right\}.$$

Similarly, $S(x, y_{j+1})$ is monotone if $S(x, y_{j+1}) > 0$ i.e.

$$S^{(1)}(x, y_{j+1}) = \frac{\sum_{i=0}^{5}(1 - \theta)^{5-i}\theta^i F_i}{(q_2(\theta))^2} > 0, \tag{8}$$

with

$$F_0 = r_{i,j+1}^2 F_{i,j+1}^x,$$
$$F_1 = 2r_{i,j+1}v_{i,j+1}\Delta_{i,j+1} - 2r_{i,j+1}\omega_{i,j+1}F_{i+1,j+1}^x + r_{i,j+1}^2 F_{i,j+1}^x,$$
$$F_2 = (3r_{i,j+1}\omega_{i,j+1} + 2r_{i,j+1}v_{i,j+1} + u_{i,j+1}v_{i,j+1})\Delta_{i,j+1} - r_{i,j+1}v_{i,j+1}F_{i,j+1}^x$$
$$\qquad -(u_{i,j+1}\omega_{i,j+1} + 2r_{i,j+1}\omega_{i,j+1})F_{i+1,j+1}^x,$$
$$F_3 = (3r_{i,j+1}\omega_{i,j+1} + 2u_{i,j+1}\omega_{i,j+1} + u_{i,j+1}v_{i,j+1})\Delta_{i,j+1} - u_{i,j+1}\omega_{i,j+1}F_{i+1,j+1}^x$$
$$\qquad -(2r_{i,j+1}\omega_{i,j+1} + r_{i,j+1}v_{i,j+1})F_{i,j+1}^x,$$
$$F_4 = 2u_{i,j+1}\omega_{i,j+1}\Delta_{i,j+1} - 2r_{i,j+1}\omega_{i,j+1}F_{i,j+1}^x + \omega_{i,j+1}^2 F_{i+1,j+1}^x,$$
$$F_5 = \omega_{i,j+1}^2 F_{i+1,j+1}^x.$$

Thus, $S^{(1)}(x, y_{j+1}) > 0$ if $\sum_{i=0}^5 (1-\theta)^{5-i}\theta^i F_i > 0$. $\sum_{i=0}^5 (1-\theta)^{5-i}\theta^i F_i > 0$ if $F_i > 0$, $i = 0, 1, 2, 3, 4, 5$. $F_i > 0$, $i = 0, 1, 2, 3, 4, 5$ if

$$u_{i,j+1} > \frac{r_{i,j+1}F_{i,j+1}^x}{\Delta_{i,j+1}},$$

$$v_{i,j+1} > Max\left\{\frac{\omega_{i,j+1}F_{i+1,j+1}^x}{\Delta_{i,j+1}}, \frac{3r_{i,j+1}\omega_{i,j+1} - u_{i,j+1}\omega_{i,j+1}F_{i+1,j+1}^x}{u_{i,j+1}\Delta_{i,j+1} + r_{i,j+1}F_{i,j+1}^x}\right\}.$$

Similarly, $S(x_i, y)$ is monotone if $S^{(1)}(x_i, y) > 0$ i.e.

$$S^{(1)}(x_i, y) = \frac{\sum_{i=0}^5 (1-\phi)^{5-i}\phi^i G_i}{(q_3(\phi))^2} > 0, \qquad (9)$$

with

$$G_0 = \hat{r}_{i,j}^2 F_{i,j}^y, \quad G_1 = 2\hat{r}_{i,j}\hat{v}_{i,j}\hat{\Delta}_{i,j} - 2\hat{r}_{i,j}\hat{\omega}_{i,j}F_{i,j+1}^y + \hat{r}_{i,j}^2 F_{i,j}^y,$$
$$G_2 = (3\hat{r}_{i,j}\hat{\omega}_{i,j} + 2\hat{r}_{i,j}\hat{v}_{i,j} + \hat{u}_{i,j}\hat{v}_{i,j})\hat{\Delta}_{i,j} - (\hat{u}_{i,j}\hat{\omega}_{i,j} + 2\hat{r}_{i,j}\hat{\omega}_{i,j})F_{i,j+1}^y - \hat{r}_{i,j}\hat{v}_{i,j}F_{i,j}^y,$$
$$G_3 = (3\hat{r}_{i,j}\hat{\omega}_{i,j} + 2\hat{u}_{i,j}\hat{\omega}_{i,j} + \hat{u}_{i,j}\hat{v}_{i,j})\hat{\Delta}_{i,j} - \hat{u}_{i,j}\hat{\omega}_{i,j}F_{i,j+1}^y - (2\hat{r}_{i,j}\hat{\omega}_{i,j} + \hat{r}_{i,j}\hat{v}_{i,j})F_{i,j}^y,$$
$$G_4 = 2\hat{u}_{i,j}\hat{\omega}_{i,j}\hat{\Delta}_{i,j} - 2\hat{r}_{i,j}\hat{\omega}_{i,j}F_{i,j}^y + \hat{\omega}_{i,j}^2 F_{i,j+1}^y, \quad G_5 = \hat{\omega}_{i,j}^2 F_{i,j+1}^y.$$

Thus, $S^{(1)}(x_i, y) > 0$ if $\sum_{i=0}^5 (1-\phi)^{5-i}\phi^i G_i > 0$. $\sum_{i=0}^5 (1-\phi)^{5-i}\phi^i G_i > 0$ if $G_i > 0$, $i = 0, 1, 2, 3, 4, 5$. $G_i > 0$, $i = 0, 1, 2, 3, 4, 5$ if

$$\hat{u}_{i,j} > \frac{\hat{r}_{i,j}F_{i,j}^y}{\hat{\Delta}_{i,j}}, \quad \hat{v}_{i,j} > Max\left\{\frac{\hat{\omega}_{i,j}F_{i,j+1}^y}{\hat{\Delta}_{i,j}}, \frac{3\hat{r}_{i,j}\hat{\omega}_{i,j} - \hat{u}_{i,j}\hat{\omega}_{i,j}F_{i,j+1}^y}{\hat{u}_{i,j}\hat{\Delta}_{i,j} + \hat{r}_{i,j}F_{i,j}^y}\right\}.$$

Similarly, $S(x_{i+1}, y)$ is monotone if $S^{(1)}(x_{i+1}, y) > 0$ i.e.

$$S^{(1)}(x_{i+1}, y) = \frac{\sum_{i=0}^5 (1-\phi)^{5-i}\phi^i H_i}{(q_4(\phi))^2} > 0, \qquad (10)$$

with

$$H_0 = \hat{r}_{i+1,j}^2 F_{i+1,j}^y,$$

$$H_1 = 2\hat{r}_{i+1,j}\hat{v}_{i+1,j}\hat{\Delta}_{i+1,j} - 2\hat{r}_{i+1,j}\hat{\omega}_{i+1,j}F_{i+1,j+1}^y + \hat{r}_{i+1,j}^2 F_{i+1,j}^y,$$

$$H_2 = (3\hat{r}_{i+1,j}\hat{\omega}_{i+1,j} + 2\hat{r}_{i+1,j}\hat{v}_{i+1,j} + \hat{u}_{i+1,j}\hat{v}_{i+1,j})\hat{\Delta}_{i+1,j} - \hat{r}_{i+1,j}\hat{v}_{i+1,j}F_{i+1,j}^y$$
$$\quad -(\hat{u}_{i+1,j}\hat{\omega}_{i+1,j} + 2\hat{r}_{i+1,j}\hat{\omega}_{i+1,j})F_{i+1,j+1}^y,$$

$$H_3 = (3\hat{r}_{i+1,j}\hat{\omega}_{i+1,j} + 2\hat{u}_{i+1,j}\hat{\omega}_{i+1,j} + \hat{u}_{i+1,j}\hat{v}_{i+1,j})\hat{\Delta}_{i+1,j} - \hat{u}_{i+1,j}\hat{\omega}_{i+1,j}F_{i+1,j+1}^y$$
$$\quad -(2\hat{r}_{i+1,j}\hat{\omega}_{i+1,j} + \hat{r}_{i+1,j}\hat{v}_{i+1,j})F_{i+1,j}^y,$$

$$H_4 = 2\hat{u}_{i+1,j}\hat{\omega}_{i+1,j}\hat{\Delta}_{i+1,j} - 2\hat{r}_{i+1,j}\hat{\omega}_{i+1,j}F_{i+1,j}^y + \hat{\omega}_{i+1,j}^2 F_{i+1,j+1}^y,$$

$$H_5 = \hat{\omega}_{i+1,j}^2 F_{i+1,j+1}^y.$$

Thus, $S^{(1)}(x_{i+1}, y) > 0$ if $\sum_{i=0}^5 (1-\phi)^{5-i}\phi^i H_i > 0$. $\sum_{i=0}^5 (1-\phi)^{5-i}\phi^i H_i > 0$ if $H_i > 0$, $i = 0, 1, 2, 3, 4, 5$. $H_i > 0$, $i = 0, 1, 2, 3, 4, 5$ if

$$\hat{u}_{i+1,j} > \frac{\hat{r}_{i+1,j}F_{i+1,j}^y}{\hat{\Delta}_{i+1,j}},$$

$$\hat{v}_{i+1,j} > Max\left\{\frac{\hat{\omega}_{i+1,j}F_{i+1,j+1}^y}{\hat{\Delta}_{i+1,j}}, \frac{3\hat{r}_{i+1,j}\hat{\omega}_{i+1,j} - \hat{u}_{i+1,j}\hat{\omega}_{i+1,j}F_{i+1,j+1}^y}{\hat{u}_{i+1,j}\hat{\Delta}_{i+1,j} + \hat{r}_{i+1,j}F_{i+1,j}^y}\right\}.$$

The above mathematical discussion can be summarized as:

**Theorem 2.** The bi-cubic partially blended rational function defined in (2) visualize monotone data in the view of monotone surface if in each rectangular patch $I_{i,j} = [x_i, x_{i+1}] \times [y_j, y_{j+1}]$, free parameters $u_{i,j}$, $v_{i,j}$, $u_{i,j+1}$, $v_{i,j+1}$, $\hat{u}_{i,j}$, $\hat{v}_{i,j}$, $\hat{u}_{i+1,j}$ and $\hat{v}_{i+1,j}$ satisfy the following conditions:

$$u_{i,j} > \frac{r_{i,j}F_{i,j}^x}{\Delta_{i,j}}, \ u_{i,j+1} > \frac{r_{i,j+1}F_{i,j+1}^x}{\Delta_{i,j+1}}, \ \hat{u}_{i,j} > \frac{\hat{r}_{i,j}F_{i,j}^y}{\hat{\Delta}_{i,j}}, \ \hat{u}_{i+1,j} > \frac{\hat{r}_{i+1,j}F_{i+1,j}^y}{\hat{\Delta}_{i+1,j}},$$

$$v_{i,j} > Max\left\{\frac{\omega_{i,j}F_{i+1,j}^x}{\Delta_{i,j}}, \frac{3r_{i,j}\omega_{i,j} - u_{i,j}\omega_{i,j}F_{i+1,j}^x}{u_{i,j}\Delta_{i,j} + r_{i,j}F_{i,j}^x}\right\},$$

$$v_{i,j+1} > Max\left\{\frac{\omega_{i,j+1}F_{i+1,j+1}^x}{\Delta_{i,j+1}}, \frac{3r_{i,j+1}\omega_{i,j+1} - u_{i,j+1}\omega_{i,j+1}F_{i+1,j+1}^x}{u_{i,j+1}\Delta_{i,j+1} + r_{i,j+1}F_{i,j+1}^x}\right\},$$

$$\hat{v}_{i,j} > Max\left\{\frac{\hat{\omega}_{i,j}F_{i,j+1}^y}{\hat{\Delta}_{i,j}}, \frac{3\hat{r}_{i,j}\hat{\omega}_{i,j} - \hat{u}_{i,j}\hat{\omega}_{i,j}F_{i,j+1}^y}{\hat{u}_{i,j}\hat{\Delta}_{i,j} + \hat{r}_{i,j}F_{i,j}^y}\right\},$$

$$\hat{v}_{i+1,j} > Max\left\{\frac{\hat{\omega}_{i+1,j}F_{i+1,j+1}^y}{\hat{\Delta}_{i+1,j}}, \frac{3\hat{r}_{i+1,j}\hat{\omega}_{i+1,j} - \hat{u}_{i+1,j}\hat{\omega}_{i+1,j}F_{i+1,j+1}^y}{\hat{u}_{i+1,j}\hat{\Delta}_{i+1,j} + \hat{r}_{i+1,j}F_{i+1,j}^y}\right\}.$$

The above constraints can be rearranged as:

$$u_{i,j} = k_{i,j} + \frac{r_{i,j}F_{i,j}^x}{\Delta_{i,j}}, \ k_{i,j} > 0, \ u_{i,j+1} = l_{i,j} + \frac{r_{i,j+1}F_{i,j+1}^x}{\Delta_{i,j+1}}, \ l_{i,j} > 0,$$

$$\hat{u}_{i,j} = m_{i,j} + \frac{\hat{r}_{i,j} F_{i,j}^y}{\hat{\Delta}_{i,j}}, \; m_{i,j} > 0, \; \hat{u}_{i+1,j} = n_{i,j} + \frac{\hat{r}_{i+1,j} F_{i+1,j}^y}{\hat{\Delta}_{i+1,j}}, \; n_{i,j} > 0,$$

$$v_{i,j} = o_{i,j} + Max \left\{ \frac{\omega_{i,j} F_{i+1,j}^x}{\Delta_{i,j}}, \; \frac{3 r_{i,j} \omega_{i,j} - u_{i,j} \omega_{i,j} F_{i+1,j}^x}{u_{i,j} \Delta_{i,j} + r_{i,j} F_{i,j}^x} \right\}, \; o_{i,j} > 0,$$

$$v_{i,j+1} = q_{i,j} + Max \left\{ \frac{\omega_{i,j+1} F_{i+1,j+1}^x}{\Delta_{i,j+1}}, \; \frac{3 r_{i,j+1} \omega_{i,j+1} - u_{i,j+1} \omega_{i,j+1} F_{i+1,j+1}^x}{u_{i,j+1} \Delta_{i,j+1} + r_{i,j+1} F_{i,j+1}^x} \right\}, \; q_{i,j} > 0,$$

$$\hat{v}_{i,j} = s_{i,j} + Max \left\{ \frac{\hat{\omega}_{i,j} F_{i,j+1}^y}{\hat{\Delta}_{i,j}}, \; \frac{3 \hat{r}_{i,j} \hat{\omega}_{i,j} - \hat{u}_{i,j} \hat{\omega}_{i,j} F_{i,j+1}^y}{\hat{u}_{i,j} \hat{\Delta}_{i,j} + \hat{r}_{i,j} F_{i,j}^y} \right\}, \; s_{i,j} > 0,$$

$$\hat{v}_{i+1,j} = t_{i,j} + Max \left\{ \frac{\hat{\omega}_{i+1,j} F_{i+1,j+1}^y}{\hat{\Delta}_{i+1,j}}, \; \frac{3 \hat{r}_{i+1,j} \hat{\omega}_{i+1,j} - \hat{u}_{i+1,j} \hat{\omega}_{i+1,j} F_{i+1,j+1}^y}{\hat{u}_{i+1,j} \hat{\Delta}_{i+1,j} + \hat{r}_{i+1,j} F_{i+1,j}^y} \right\}, \; t_{i,j} > 0.$$

$r_{i,j}$, $\omega_{i,j}$, $\hat{r}_{i,j}$ and $\hat{\omega}_{i,j}$ are positive real numbers.

## 5  Demonstration

Let us consider a 3D monotone data in Table 1. Fig.1(a) is produced from the monotone data set in Table 1 using bi-cubic Hermite spline which looses the shape of data. Monotone surface in Fig.1(b) is produced from the same data set using monotone surface data interpolation scheme developed in Section 4 with $r_{i,j} = 0.001$, $\hat{r}_{i,j} = 0.004$, $\omega_{i,j} = 0.003$, $\hat{\omega}_{i,j} = 0.005$. The surface in Fig.1(b) is monotone but not smooth. The monotone surface is further refined in Fig.2(a) and (b) with ($r_{i,j} = 0.02$, $\hat{r}_{i,j} = 0.04$, $\omega_{i,j} = 0.03$, $\hat{\omega}_{i,j} = 0.05$) and ($r_{i,j} = 1.2$, $\hat{r}_{i,j} = 1$, $\omega_{i,j} = 1.2$, $\hat{\omega}_{i,j} = 1$) respectively.

**Table 1.** A 3D monotone data

| y/x | 11 | 12 | 14 | 15 |
|-----|----|----|----|----|
| 11 | 15 | 15.1 | 15.2 | 15.3 |
| 12 | 56 | 56.1 | 56.2 | 56.3 |
| 14 | 60 | 60.1 | 60.2 | 60.3 |
| 15 | 85 | 85.1 | 85.2 | 85.3 |

Another example is presented here for a monotone data set (up to four decimal places) shown in Table 2. Fig.3(a) is a demonstration to this data using bi-cubic Hermite spline which looses the shape of the data. Monotone surface, in Fig.3(b), is produced from the same data set using monotone surface data interpolation scheme developed in Section 4 with $r_{i,j} = 0.0005$, $\hat{r}_{i,j} = 0.0001$, $\omega_{i,j} = 0.0006$, $\hat{\omega}_{i,j} = 0.0003$. This surface may be further refined and enhanced, if needed, using the free parameters in the scheme. This refinement has been demonstrated in Fig.4(a) and (b) with ($r_{i,j} = 0.05$, $\hat{r}_{i,j} = 0.01$, $\omega_{i,j} = 0.06$, $\hat{\omega}_{i,j} = 0.03$) and ($r_{i,j} = 1.2$, $\hat{r}_{i,j} = 1$, $\omega_{i,j} = 1.2$, $\hat{\omega}_{i,j} = 1$) respectively.

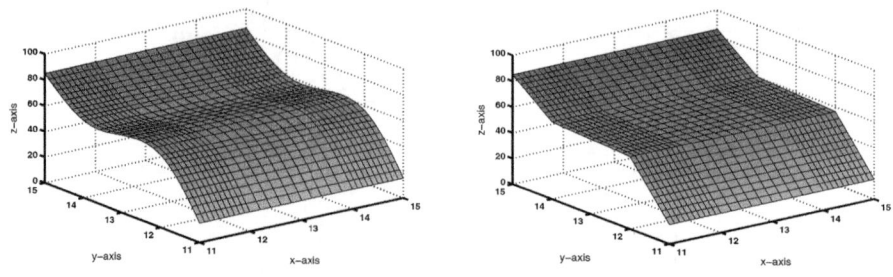

**Fig. 1.** (a) Bi-cubic Hermite spline. (b) Monotone bi-cubic partially blended rational function.

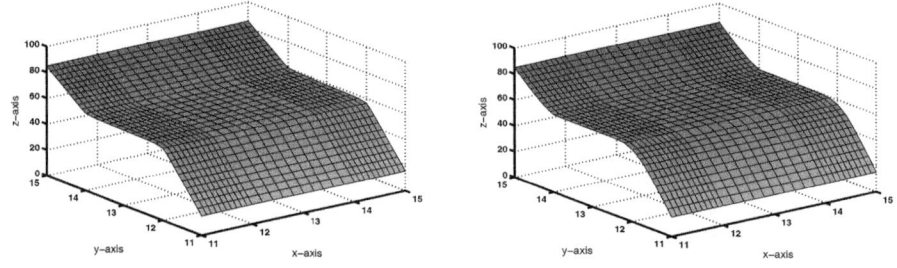

**Fig. 2.** (a) Monotone bi-cubic partially blended rational function. (b) Monotone bi-cubic partially blended rational function.

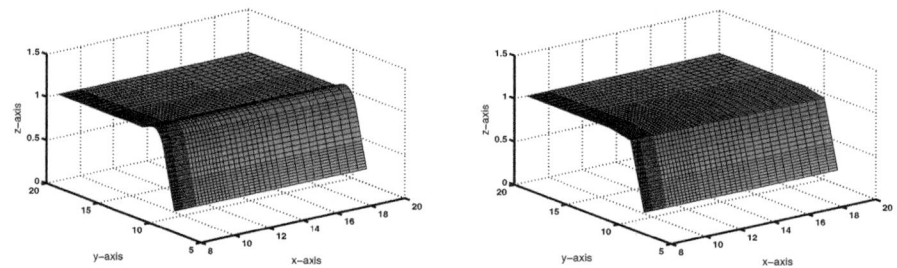

**Fig. 3.** (a) Bi-cubic Hermite spline. (b) Monotone bi-cubic partially blended rational function.

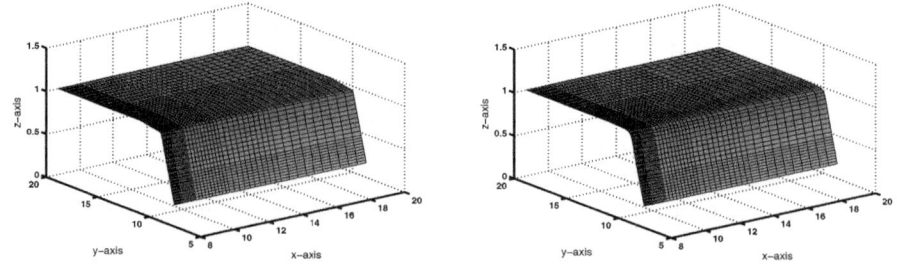

**Fig. 4.** (a) Monotone bi-cubic partially blended rational function. (b) Monotone bi-cubic partially blended rational function.

**Table 2.** Another 3D monotone data with four decimal entries

| y/x | 8.7 | 9.2 | 10 | 12 | 15 | 20 |
|---|---|---|---|---|---|---|
| **8.7** | 0.1691 | 0.1692 | 0.1693 | 0.1693 | 0.1694 | 0.1694 |
| **9.2** | 0.4694 | 0.4695 | 0.4696 | 0.4697 | 0.4698 | 0.4699 |
| **10** | 0.9437 | 0.9438 | 0.9439 | 0.9440 | 0.9441 | 0.9442 |
| **12** | 0.9986 | 0.9987 | 0.9988 | 0.9989 | 0.9990 | 0.9991 |
| **15** | 0.9994 | 0.9995 | 0.9996 | 0.9997 | 0.9998 | 0.9999 |
| **20** | 1.0001 | 1.0002 | 1.0003 | 1.0004 | 1.0005 | 1.0006 |

# 6 Conclusion

A partially blended rational bi-cubic function, with eight families of parameters, has been developed. Out of these, four shape parameters are constrained to preserve the monotonicity, while the other four parameters have been left as free. The free parameters are used to refine the shape of the surfaces. The constraints are data dependent to preserve the shape of monotone data. The proposed scheme applies equally well to data or data with given derivatives. It is more flexible and suitable for interactive CAD systems.

# References

1. Beatson, R.K., Ziegler, Z.: Monotonicity preserving surface interpolation. SIAM Journal of Numerical Analysis 22(2), 401–411 (1985)
2. Carlson, R.E., Fritsch, F.N.: Monotone piecewise bicubic interpolation. SIAM Journal of Numerical Analysis 22, 386–400 (1985)
3. Casciola, G., Romani, L.: Rational interpolants with tension parameters. In: Lyche, T., Mazure, M., Schumaker, L.L. (eds.) Proceedings of Curve and Surface Design, Saint-Malo 2002, pp. 41–50. Nashboro Press, Brentwood (2003)
4. Clemens, P., Jütter, B.: Monotonicity-preserving interproximation of B-H- curves. Journal of Computational and Applied Mathematics 196(1), 45–57 (2006)
5. Costantini, P., Fontanella, F.: Shape preserving bivariate interpolation. SIAM Journal of Numerical Analysis 27(2), 488–506 (1990)
6. Floater, M.S., Peña, J.M.: Monotonicity preservation on triangles. Mathematics of Computation 69(232), 1505–1519 (2000)
7. Han, L., Schumaker, L.L.: Fitting monotone surfaces to scattered data using $C^1$ piecewise cubics. SIAM Journal of Numerical Analysis 23(2), 569–585 (1997)
8. Hussain, M.Z., Hussain, M.: Visualization of data preserving monotonicity. Applied Mathematics and Computation 190, 1353–1364 (2007)
9. Hussain, M.Z., Sarfarz, M.: Monotone piecewise rational cubic interpolation. International Journal of Computer Mathematics 86(3), 423–430 (2009)
10. Sarfraz, M., Butt, S., Hussain, M.Z.: Surfaces for the visualization of scientific data preserving monotonicity. In: Proceedings of the IMA Mathematics for Surfaces VII Conference, Dundee, UK, September 2-5, pp. 479–495 (1997)

# $C^1$ Monotone Scattered Data Interpolation

Malik Zawwar Hussain and Maria Hussain

Department of Mathematics
University of the Punjab
Lahore-Pakistan
malikzawwar@math.pu.edu.pk

**Abstract.** A local $C^1$ surface construction scheme is presented to preserve the shape of the monotone scattered data arranged over the triangular grid. Each boundary curve of the triangle is constructed by the rational cubic function and this rational function is also used for the side-vertex interpolation. The final surface patch is constructed by taking the convex combination of three side-vertex interpolants. For each triangular patch there are three boundary curves and three side vertex interpolant. Simple sufficient data dependent constraints are derived on the free parameters in the description of rational function to preserve the shape of monotone scattered data. The developed scheme is local, computationally economical and visually pleasing.

**Keywords:** Triangular surface patch, Rational cubic function, Free parameters, Monotone scattered data.

## 1 Introduction

In Computer Graphics environment, a user is always in need of interpolatory scheme which possesses certain characteristics like shape preservation, shape control, etc. Ordinary interpolation techniques although smoother are not helpful for the visualization of shaped data. Positive, monotone and convex are the fundamental shapes of data. Monotonicity is an important shape property. There are many physical situations where entities only have a meaning when their values are monotone e.g., dose-response curve and surfaces in biochemistry and pharmacology [2], approximations of couples and quasi couples in statistics [2], empirical option pricing model in finance [2], approximation of potential functions in physical and chemical system [2].

The problem of shape preservation of monotone scattered data has been discussed by a number of authors. Beliakov [2] presented a method of monotone interpolation and smoothing of multivariate scattered data. In [2] smoothing of noisy data subject to monotonicity constraints was converted into a quadratic programming problem. The scheme developed in [2] was only applicable to preserve the shape of monotone Lipschitz continuous functions. Delgado and Peña [4] proved that rational Bézier surfaces on rectangular and triangular grid are not axially monotonicity preserving. Floater and Peña [6] defined three types of monotonicity preservation for the systems of bivariate functions on a triangle.

M.L. Gavrilova et al. (Eds.): Trans. on Comput. Sci. VIII, LNCS 6260, pp. 156–166, 2010.

The authors in [8] to preserve the shape of monotone scattered data, first converted the given scattered data into a rectangular grid by drawing horizontal and vertical lines through data sites and the functional values at the new data sites were calculated by an interpolant that interpolates the original scattered data. The draw back of the method was that a system of $N-$scattered data points reduced to $N^2-$rectangles and some of the rectangles were very small in one or both directions.

This paper is particularly concerned with the construction of interpolating surfaces that preserve the monotonicity of scattered data in the given direction using a local side-vertex method that is applicable to both data with and without derivatives. The rational cubic function [11] is used for boundary as well as for side-vertex interpolation, yielding twelve free parameters in each triangular patch. Simple sufficient data dependent constraints are derived on these free parameters to preserve the shape of monotone scattered data.

The remainder of the paper is organized as follows: Section 2 reviews the side-vertex method for interpolation over a triangle. The problem of monotonicity is discussed in Section 3, where the rational cubic function [11] is used for boundary as well as for side-vertex interpolation. The surface scheme of Section 3 is demonstrated in Section 4. Finally, Section 5 discusses the result from Section 3 and concludes the paper.

## 2 $C^1$ Side-Vertex Method for Interpolation over a Triangle

In this Section, we shall review the $C^1$ side-vertex method proposed by Nielson [9] to interpolate the data arranged over the triangular grid to generate the triangular patches.

For a non-degenerate triangle $T$, with vertices $\{V_i = (x_i, y_i), i = 1, 2, 3\}$, barycentric coordinates $u$, $v$ and $w$, any point $V = (x, y)$ on the triangle can be expressed as:

$$V = uV_1 + vV_2 + wV_3, \ u + v + w = 1, \ u, \ v, \ w \geq 0. \tag{1}$$

The Nielson $C^1$ side-vertex interpolant over each triangular patch is defined by the following convex combination:

$$P(u, v, w) = \frac{v^2 w^2 P_1 + u^2 w^2 P_2 + u^2 v^2 P_3}{v^2 w^2 + u^2 w^2 + u^2 v^2}, \tag{2}$$

where $P_i$, $i = 1, 2, 3$ are the radial curves connecting vertices $V_i$, $i = 1, 2, 3$ with the points $S_i$, $i = 1, 2, 3$ on the opposite boundary edges $e_i$, $i = 1, 2, 3$.

Nielson [9] used the cubic Hermite interpolant to define the three boundary curves along the three edges of the triangle as well as radial curves $P_i$, $i = 1, 2, 3$ connecting the vertices $V_i$, $i = 1, 2, 3$ to the points $S_i$, $i = 1, 2, 3$ on the opposite boundary edges $e_i$, $i = 1, 2, 3$.

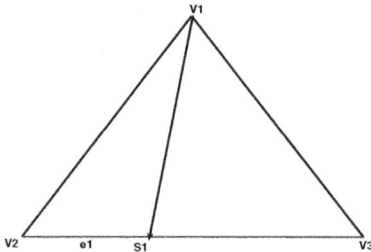

**Fig. 1.** Locations of the vertices and edges of the triangle $\triangle V_1V_2V_3$

The radial curve $P_1$ connecting the vertex $V_1$ to the point $S_1$ on the opposite edge $e_1$ is defined as:

$$P_1 = B_1 + B_2, \tag{3}$$

where

$B_1 = (1-u)^2(1+2u)F(S_1) - u(1-u)\left\{v\frac{\partial F}{\partial e_3}(S_1) + w\frac{\partial F}{\partial e_2}(S_1)\right\}$,

$B_2 = u^2(3-2u)F(V_1) + u^2\left\{v\frac{\partial F}{\partial e_3}(V_1) + w\frac{\partial F}{\partial e_2}(V_1)\right\}$,

with

$\frac{\partial F}{\partial e_2}(V_1) = (x_3-x_1)\frac{\partial F}{\partial x}(V_1) + (y_3-y_1)\frac{\partial F}{\partial y}(V_1)$, $\frac{\partial F}{\partial e_3}(V_1) = (x_2-x_1)\frac{\partial F}{\partial x}(V_1) + (y_2-y_1)\frac{\partial F}{\partial y}(V_1)$,

$\frac{\partial F}{\partial e_2}(S_1) = (x_3-x_1)\frac{\partial F}{\partial x}(S_1) + (y_3-y_1)\frac{\partial F}{\partial y}(S_1)$, $\frac{\partial F}{\partial e_3}(S_1) = (x_2-x_1)\frac{\partial F}{\partial x}(S_1) + (y_2-y_1)\frac{\partial F}{\partial y}(S_1)$.

$F(S_1)$ is the boundary curve(cubic Hermite interpolant) along the edge $e_1$ defined as

$$F(S_1) = r^3 F_2 + r^2 r_1 (F_2 + D_2) + rr_1^2(F_3 - D_3) + r_1^3 F_3, \tag{4}$$

$r = \frac{v}{v+w}$, $r_1 = \frac{w}{v+w}$. $(F_i, D_i)$, $i = 2,3$ are the functional and derivative values at the vertices $V_2$ and $V_3$. The radial curves $P_2$ and $P_3$ connecting the vertices $V_i$, $i = 2,3$ to the points $S_i$, $i = 2,3$ on the opposite boundary edges $e_i$, $i = 2,3$ are defined in the similar way. From (3)-(4) it is clear that the interpolant (2), not only interpolates the data values at the vertices but the first order derivatives at the boundary edges as well.

At the vertices of the triangle $\triangle V_1V_2V_3$ where two of the barycentric coordinates are simultaneously zero, the interpolant (2) is defined as: $P(u,0,0) = F_1$, $P(0,v,0) = F_2$, $P(0,0,w) = F_3$, where $F_i$, $i = 1,2,3$ are the ordinates values at the vertices $V_i = (x_i, y_i)$, $i = 1,2,3$.

## 3    The $C^1$ Monotonicity Preserving Scattered Data Interpolation

Let $\{(x_i, y_i, F_i), \ i = 1,2,3\}$ be the monotone scattered data defined over the triangle $\triangle V_1V_2V_3$ i.e.

$$F_i < F_j \ for \ x_i < x_j \ and \ y_i < y_j, \ F_i^x > 0 \ and \ F_i^y > 0, \ i = 1, 2, 3.$$

Michael S. Floater and J. M. Peña [6] defined the monotonicity preservation in a certain direction $d = \lambda_1 V_1 + \lambda_2 V_2 + \lambda_3 V_3$, $\lambda_1 + \lambda_2 + \lambda_3 = 0$ as:

**Definition 1.** A function $f(x, y)$ is said to be strictly monotonic in any direction $d$ if

$$D_d f(x, y) > 0,$$

where $D_d$ denotes the directional derivative along the direction $d$. □

## 3.1 Triangulation

The domain is triangulated by the Delaunay Triangulation method [5] such that the data $F_i$ are lying at the vertices $V_i = (x_i, y_i)$, $i = 1, 2, 3, ..., n$ of the triangles.

## 3.2 Estimation of Derivatives

Partial derivatives at the vertices $V_i$, $i = 1, 2, 3$ are estimated by derivative estimation technique proposed in [7].

## 3.3 The $C^1$ Monotone Triangular Patch

Let $\triangle V_1 V_2 V_3$ be the given triangle having edges $e_i$, $i = 1, 2, 3$ opposite to the vertices $V_i$, $i = 1, 2, 3$ respectively.

The radial curve $P_1$ connecting a vertex $V_1$ to the point $S_1$ on the opposite edge $e_1$ is defined as:

$$P_1 = \frac{P_{1n}}{P_{1d}}, \tag{5}$$

where

$$P_{1n} = \{u^3 \alpha_i + (1-u)u^2 A_1\}F_1 + \{(1-u)^3 \beta_i + (1-u)^2 u A_2\}F(S_1) + u^2 v \alpha_i R_3$$
$$+ u^2 w \alpha_i R_4 - u(1-u)v\beta_i R_1 - u(1-u)w\beta_i R_2,$$
$$P_{1d} = \alpha_i u^2 + 2\alpha_i \beta_i (1-u)u + \beta_i(1-u)^2,$$

$$R_1 = (x_2 - x_1)\frac{\partial F}{\partial x}(S_1) + (y_2 - y_1)\frac{\partial F}{\partial y}(S_1), \ R_2 = (x_3 - x_1)\frac{\partial F}{\partial x}(S_1) + (y_3 - y_1)\frac{\partial F}{\partial y}(S_1),$$
$$R_3 = (x_2 - x_1)\frac{\partial F}{\partial x}(V_1) + (y_2 - y_1)\frac{\partial F}{\partial y}(V_1), \ R_4 = (x_3 - x_1)\frac{\partial F}{\partial x}(V_1) + (y_3 - y_1)\frac{\partial F}{\partial y}(V_1),$$
$$d_3 = (x_3 - x_2)\frac{\partial F}{\partial x}(V_2) + (y_3 - y_2)\frac{\partial F}{\partial y}(V_2), \ d_4 = (x_3 - x_2)\frac{\partial F}{\partial x}(V_3) + (y_3 - y_2)\frac{\partial F}{\partial y}(V_3),$$
$$A_1 = 2\alpha_i \beta_i + \alpha_i, \ A_2 = 2\alpha_i \beta_i + \beta_i, \ r = \frac{v}{v+w}, \ r_1 = \frac{w}{v+w}.$$

Similarly, the radial curve $P_2$ and $P_3$ connecting the vertices $V_2$ and $V_3$ to the points $S_2$ and $S_3$ on the opposite edges $e_2$ and $e_3$ are defined as:

$$P_2 = \frac{P_{2n}}{P_{2d}}, \tag{6}$$

where

$$P_{2n} = \{v^3\alpha_j + (1-v)v^2 B_1\}F_2 + \{(1-v)^3\beta_j + (1-v)^2 vB_2\}F(S_2) + v^2 u\alpha_j R_7$$
$$+ v^2 w\alpha_j R_8 - v(1-v)u\beta_j R_5 - v(1-v)w\beta_j R_6,$$
$$P_{2d} = \alpha_j v^2 + 2\alpha_j\beta_j(1-v)v + \beta_j(1-v)^2,$$

$$P_3 = \frac{P_{3n}}{P_{3d}}, \tag{7}$$

where

$$P_{3n} = \{w^3\alpha_k + (1-w)w^2 C_1\}F_3 + \{(1-w)^3\beta_k + (1-w)^2 wC_2\}F(S_3) + w^2 u\alpha_k R_{11}$$
$$+ w^2 v\alpha_k R_{12} - w(1-w)u\beta_k R_9 - w(1-w)v\beta_k R_{10},$$
$$P_{3d} = \alpha_k w^2 + 2\alpha_k\beta_k(1-w)w + \beta_k(1-w)^2,$$

$F(S_2) = \frac{F(S_2)_N}{F(S_2)_D}$, $F(S_3) = \frac{F(S_3)_N}{F(S_3)_D}$,

$F(S_2)_N = \gamma_2 F_3 s_1^3 + \{(2\gamma_2\eta_2 + \gamma_2)F_3 + \gamma_2 d_5\}s_1^2 s + \{(2\gamma_2\eta_2 + \eta_2)F_1 - \eta_2 d_6\}s_1 s^2 + \eta_2 F_1 s^3$,

$F(S_2)_D = \gamma_2 s_1^2 + 2\gamma_2\eta_2 s_1 s + \eta_2 s^2$,

$F(S_3)_N = \gamma_3 F_1 t_1^3 + \{(2\gamma_3\eta_3 + \gamma_3)F_1 + \gamma_3 d_1\}t_1^2 t + \{(2\gamma_3\eta_3 + \eta_3)F_2 - \eta_3 d_2\}t_1 t^2 + \eta_3 F_2 t^3$,

$F(S_3)_D = \gamma_3 t_1^2 + 2\gamma_3\eta_3 t_1 t + \eta_3 t^2$,

$R_5 = (x_1 - x_2)\frac{\partial F}{\partial x}(S_2) + (y_1 - y_2)\frac{\partial F}{\partial y}(S_2)$, $R_6 = (x_3 - x_2)\frac{\partial F}{\partial x}(S_2) + (y_3 - y_2)\frac{\partial F}{\partial y}(S_2)$,

$R_7 = (x_1 - x_2)\frac{\partial F}{\partial x}(V_2) + (y_1 - y_2)\frac{\partial F}{\partial y}(V_2)$, $R_8 = (x_3 - x_2)\frac{\partial F}{\partial x}(V_2) + (y_3 - y_2)\frac{\partial F}{\partial y}(V_2)$,

$R_9 = (x_2 - x_3)\frac{\partial F}{\partial x}(V_3) + (y_2 - y_3)\frac{\partial F}{\partial y}(V_3)$, $R_{10} = (x_1 - x_3)\frac{\partial F}{\partial x}(V_3) + (y_1 - y_3)\frac{\partial F}{\partial y}(V_3)$,

$R_{11} = (x_1 - x_3)\frac{\partial F}{\partial x}(S_3) + (y_1 - y_3)\frac{\partial F}{\partial y}(S_3)$, $R_{12} = (x_2 - x_3)\frac{\partial F}{\partial x}(S_3) + (y_2 - y_3)\frac{\partial F}{\partial y}(S_3)$,

$d_5 = (x_1 - x_3)\frac{\partial F}{\partial x}(V_3) + (y_1 - y_3)\frac{\partial F}{\partial y}(V_3)$, $d_6 = (x_1 - x_3)\frac{\partial F}{\partial x}(V_1) + (y_1 - y_3)\frac{\partial F}{\partial y}(V_1)$,

$d_1 = (x_2 - x_1)\frac{\partial F}{\partial x}(V_1) + (y_2 - y_1)\frac{\partial F}{\partial y}(V_1)$, $d_2 = (x_2 - x_1)\frac{\partial F}{\partial x}(V_2) + (y_2 - y_1)\frac{\partial F}{\partial y}(V_2)$,

$B_1 = 2\alpha_j\beta_j + \alpha_j$, $B_2 = 2\alpha_j\beta_j + \beta_j$, $C_1 = 2\alpha_k\beta_k + \alpha_k$, $C_2 = 2\alpha_k\beta_k + \beta_k$,

$s_1 = \frac{w}{u+w}$, $s = \frac{u}{u+w}$, $t_1 = \frac{u}{u+v}$, $t = \frac{v}{u+v}$.

The directional derivative of (2) along the direction $d = \lambda_1 V_1 + \lambda_2 V_2 + \lambda_3 V_3$, with $\lambda_1 + \lambda_2 + \lambda_3 = 0$ is

$$D_d P = \lambda_1 \frac{\partial P}{\partial u} + \lambda_2 \frac{\partial P}{\partial v} + \lambda_3 \frac{\partial P}{\partial w} = \frac{(D_d P)_N}{(D_d P)_D}, \tag{8}$$

$$(D_d P)_N = u^2 v^4 w^2 E_1 + u^4 v^2 w^2 E_2 + u^4 v^4 E_3 + v^4 w^4 E_4 + uv^2 w^4 E_5 + u^2 v^2 w^4 E_6$$
$$+ uv^4 w^2 E_7 + u^4 w^4 E_8 + u^2 vw^4 E_9 + u^2 v^4 w E_{10} + u^4 vw^2 E_{11}$$
$$+ u^4 v^2 w E_{12} \tag{9}$$

$$(D_d P)_D = (u^2 v^2 + v^2 w^2 + w^2 u^2)^2. \tag{10}$$

From **Definition 1**, (2) is monotone if $D_d P > 0$. From (8), $D_d P > 0$ if $(D_d P)_N > 0$, $(D_d P)_D > 0$. From (9), $(D_d P)_N > 0$ if $E_i > 0$, $i = 1, 2, 3, ..., 12$.

From (10), $(D_d P)_D > 0$ is always true as $0 \leq u, v, w \leq 1$. $E_i > 0$, $i = 1, 2, 3, ..., 12$ if

$$\eta_1 > Max\{0, \ Con_l, \ 1 \leq l \leq 3\}, \ \eta_2 > Max\{0, \ Con_l, \ 4 \leq l \leq 6\},$$
$$\eta_3 > Max\{0, \ Con_l, \ 7 \leq l \leq 9\}, \ \gamma_1 > Max\{0, \ Con_l, \ 10 \leq l \leq 12\},$$
$$\gamma_2 > Max\{0, \ Con_l, \ 13 \leq l \leq 15\}, \ \gamma_3 > Max\{0, \ Con_l, \ 16 \leq l \leq 18\},$$
$$\alpha_i > Max\{0, \ Con_l, \ 19 \leq l \leq 25\}, \ \alpha_j > Max\{0, \ Con_l, \ 26 \leq l \leq 32\},$$
$$\alpha_k > Max\{0, \ Con_l, \ 33 \leq l \leq 39\},$$

$$Con_1 = \frac{\lambda_2 d_3}{4\lambda_3 (F_3 - F_2)}, \ Con_2 = \frac{-d_3}{2(F_2 - F_3)},$$

$$Con_3 = \frac{(-4\lambda_2 + 2\lambda_3)(F_2 - F_3) + (-\lambda_2 + 2\lambda_3)d_3 - \lambda_2 d_4}{(2\lambda_2 - 4\lambda_3)(F_2 - F_3) + (2\lambda_2 - \lambda_3)d_4},$$

$$Con_4 = \frac{\lambda_3 d_5}{4\lambda_1 (F_3 - F_1)}, \ Con_5 = \frac{-d_5}{2(F_3 - F_1)},$$

$$Con_6 = \frac{-(4\lambda_3 - 2\lambda_1)(F_3 - F_1) - (\lambda_3 - 2\lambda_1)d_5 - \lambda_3 d_6}{(2\lambda_3 - 4\lambda_1)(F_3 - F_1) + (2\lambda_3 - \lambda_1)d_6},$$

$$Con_7 = \frac{\lambda_1 d_1}{4\lambda_2 (F_2 - F_1)}, \ Con_8 = \frac{-d_1}{2(F_1 - F_2)},$$

$$Con_9 = \frac{-(4\lambda_1 - 2\lambda_2)(F_1 - F_2) - (\lambda_1 - 2\lambda_2)d_1 - \lambda_1 d_2}{(2\lambda_1 - 4\lambda_2)(F_1 - F_2) + (2\lambda_1 - \lambda_2)d_2},$$

$$Con_{10} = \frac{-\lambda_3 \beta_1 d_4}{\lambda_2 (4\beta_1 + 2)(F_2 - F_3) + 2\lambda_2 d_3}, \ Con_{11} = \frac{-2\beta_1 \lambda_3 (F_3 - F_2 - d_4)}{4\beta_1 \lambda_3 (F_3 - F_2) - \lambda_2 d_3},$$

$$Con_{12} = \frac{-(2\lambda_2 - 4\lambda_3)(F_2 - F_3) - (2\lambda_2 - \lambda_3)d_4 + \lambda_3 d_3}{(4\lambda_2 - 2\lambda_3)(F_2 - F_3) + (\lambda_2 - 2\lambda_3)d_3},$$

$$Con_{13} = \frac{-\lambda_1 \beta_2 d_6}{\lambda_3 (4\beta_2 + 2)(F_3 - F_1) + 2\lambda_3 d_5}, \ Con_{14} = \frac{-2\beta_2 \lambda_1 (F_1 - F_3 - d_6)}{4\beta_2 \lambda_1 (F_1 - F_3) - \lambda_3 d_5},$$

$$Con_{15} = \frac{-(4\lambda_1 - 2\lambda_3)(F_1 - F_3) + (\lambda_1 - 2\lambda_3)d_6 + \lambda_1 d_5}{(2\lambda_1 - 4\lambda_3)(F_1 - F_3) + (\lambda_3 - 2\lambda_1)d_5},$$

$$Con_{16} = \frac{-\lambda_2 \beta_3 d_2}{\lambda_1 (4\beta_3 + 2)(F_1 - F_2) + 2\lambda_1 d_1}, \ Con_{17} = \frac{-2\lambda_2 \beta_3 (F_2 - F_1 - d_2)}{4\beta_3 \lambda_2 (F_2 - F_1) - \lambda_1 d_1},$$

$$Con_{18} = \frac{-(4\lambda_2 - 2\lambda_1)(F_2 - F_1) + (\lambda_2 - 2\lambda_1)d_2 + \lambda_2 d_1}{(2\lambda_2 - 4\lambda_1)(F_2 - F_1) + (\lambda_1 - 2\lambda_2)d_1},$$

$$Con_{19} = \frac{-R_1}{2R_3}, \ Con_{20} = \frac{-R_2}{2R_4}, \ Con_{21} = \frac{\beta_i R_1}{2R_3}, \ Con_{22} = \frac{\beta_i R_2}{2R_4},$$

$$Con_{23} = \frac{2\lambda_1 (F(S_1) - F_1) + \lambda_2 R_1 + \lambda_3 R_2}{4\lambda_1 (F_1 - F(S_1)) + 2(\lambda_2 R_3 + \lambda_3 R_4)},$$

$$Con_{24} = \frac{-(2\beta_i + 4)\lambda_1 (F_1 - F(S_1)) + 2\beta_i (\lambda_2 R_1 + \lambda_3 R_2) - (\lambda_2 R_3 + \lambda_3 R_4)}{\lambda_1 (4\beta_i + 2)(F_1 - F(S_1))},$$

$$Con_{25} = \frac{\beta_i (\lambda_2 R_1 + \lambda_3 R_2)}{\lambda_1 (-4\beta_i - 2)(F(S_1) - F_1)}, \ Con_{26} = \frac{-R_5}{2R_7}, \ Con_{27} = \frac{-R_6}{2R_8},$$

$$Con_{28} = \frac{\beta_j R_5}{2R_7}, \quad Con_{29} = \frac{\beta_j R_6}{2R_8}, \quad Con_{30} = \frac{2\lambda_2(F(S_2) - F_2) + \lambda_1 R_5 + \lambda_3 R_6}{4\lambda_2(F_2 - F(S_2)) + 2(\lambda_1 R_7 + \lambda_3 R_8)},$$

$$Con_{31} = \frac{-(2\beta_j + 4)\lambda_2(F_2 - F(S_2)) + 2\beta_j(\lambda_1 R_5 + \lambda_3 R_6) - (\lambda_1 R_7 + \lambda_3 R_8)}{\lambda_2(4\beta_j + 2)(F_2 - F(S_2))},$$

$$Con_{32} = \frac{\beta_j(\lambda_1 R_5 + \lambda_3 R_6)}{\lambda_2(4\beta_j + 2)(-F(S_2) + F_2)}, \quad Con_{33} = \frac{-R_9}{2R_{11}}, \quad Con_{34} = \frac{-R_{10}}{2R_{12}},$$

$$Con_{35} = \frac{\beta_k R_9}{2R_{11}}, \quad Con_{36} = \frac{\beta_k R_{10}}{2R_{12}}, \quad Con_{37} = \frac{2\lambda_3(F(S_3) - F_3) + \lambda_1 R_9 + \lambda_2 R_{10}}{4\lambda_3(F_3 - F(S_3)) + 2(\lambda_1 R_{11} + \lambda_2 R_{12})},$$

$$Con_{38} = \frac{(2\beta_k + 4)\lambda_3(-F_3 + F(S_3)) + 2\beta_k(\lambda_1 R_9 + \lambda_2 R_{10}) - (\lambda_1 R_{11} + \lambda_2 R_{12})}{\lambda_3(-4\beta_k - 2)(-F_3 + F(S_3))},$$

$$Con_{39} = \frac{\beta_k(\lambda_1 R_9 + \lambda_2 R_{10})}{\lambda_3(-4\beta_k - 2)(-F_3 + F(S_3))}.$$

The above can be summarized as:

**Theorem 1.** The $C^1$ triangular patch $P$ , defined over the triangular domain, in (2), is monotone in the direction $d = \lambda_1 V_1 + \lambda_2 V_2 + \lambda_3 V_3$, with $\lambda_1 + \lambda_2 + \lambda_3 = 0$ if the following sufficient conditions are satisfied:(after rearrangement of constraints written on Page 161.)

$\beta_i > 0, \ \beta_j > 0, \ \beta_k > 0,$

$\eta_1 = g_1 + Max\{0, \ Con_l, \ 1 \le l \le 3\}, \ \eta_2 = g_2 + Max\{0, \ Con_l, \ 4 \le l \le 6\},$

$\eta_3 = g_3 + Max\{0, \ Con_l, \ 7 \le l \le 9\}, \ \gamma_1 = g_4 + Max\{0, \ Con_l, \ 10 \le l \le 12\},$

$\gamma_2 = g_5 + Max\{0, \ Con_l, \ 13 \le l \le 15\}, \ \gamma_3 = g_6 + Max\{0, \ Con_l, \ 16 \le l \le 18\},$

$\alpha_i = g_7 + Max\{0, \ Con_l, \ 19 \le l \le 25\}, \ \alpha_j = g_8 + Max\{0, \ Con_l, \ 26 \le l \le 32\},$

$\alpha_k = g_9 + Max\{0, \ Con_l, \ 33 \le l \le 39\},$

where $g_i > 0$, $i = 1, 2, ..., 9$. The $Con_l$, $1 \le l \le 39$ are already defined in Section 3.

## 4    Demonstration

In this section, we shall illustrate our monotonicity preserving interpolating scheme developed in Section 3 using two test functions.

**Example 1:** We take the following Yager triangular norm function from [2]:

$$F_1(x, y) = max(0, \ 1 - \sqrt{(1 - x)^2 + (1 - y)^2}).$$

The monotone data points are generated from the above function with the restriction of domain $[0, 1] \times [0, 1]$. Figure 2 is the Delaunay triangulation of domain. Figure 3 is the linear interpolation of scattered data generated from the function $F_1(x, y)$. Figure 4 is the monotone surface generated from Theorem 1 with $g_i = 0.45$, $i = 1, ..., 9$, $\beta_i = 0.3$, $\beta_j = 0.39$ and $\beta_k = 0.35$ in the direction

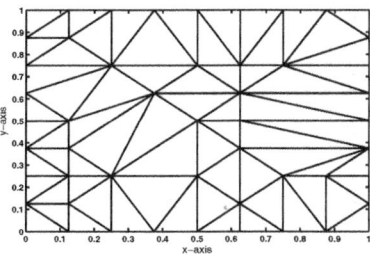

**Fig. 2.** Triangulation of domain for $F_1(x, y)$

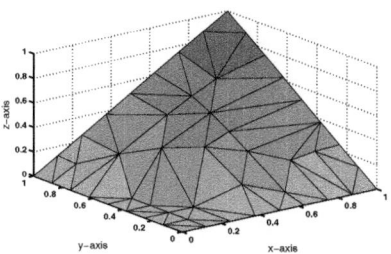

**Fig. 3.** Linear interpolation of the data generated from $F_1(x, y)$

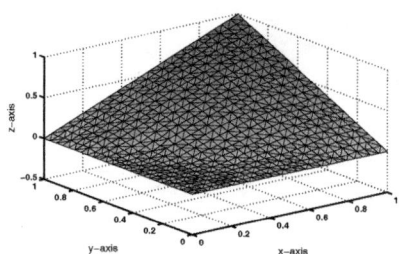

**Fig. 4.** The monotone surface generated from the Theorem 1 with $g_i = 0.45$, $i = 1, ..., 9$, $\beta_i = 0.3$, $\beta_j = 0.39$ and $\beta_k = 0.35$

$d = (1, 1)$. From Figure 4 it is clear that the shape of monotone scattered data has been preserved and it is visually pleasing as well.

**Example 2:** Monotone data for the second example is generated from the following 3-$\prod$ uniform function taken from [2]:

$$F_2(x, y) = \frac{xy}{xy + (1 - x)(1 - y)}.$$

The monotone data points are generated from the above function with the restriction of domain $[0, 0.875] \times [0, 0.875]$.

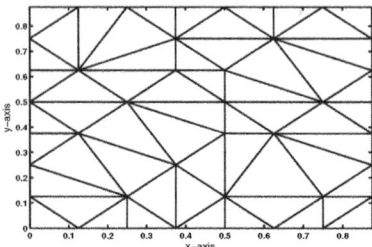

**Fig. 5.** Triangulation of domain for $F_2(x, y)$

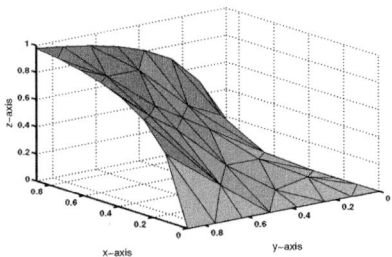

**Fig. 6.** Linear interpolation of the data generated from $F_2(x, y)$

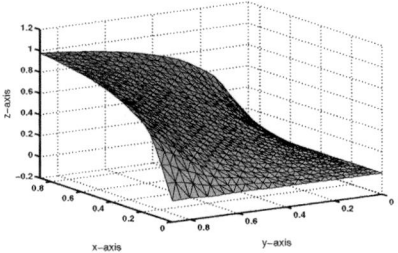

**Fig. 7.** The monotone surface generated from the Theorem 1 with $g_i = 0.6$, $i = 1, ..., 9$, $\beta_i = 0.35$, $\beta_j = 0.3$ and $\beta_k = 0.4$

Figure 5 is the Delaunay triangulation of domain. Figure 6 is the linear interpolation of scattered data generated from the function $F_2(x, y)$. Figure 7 is the monotone surface generated from Theorem 1 with $g_i = 0.6$, $i = 1, ..., 9$, $\beta_i = 0.35$, $\beta_j = 0.3$ and $\beta_k = 0.4$ in the direction $d = (1, 0)$. From Figure 7 it is clear that the shape of monotone scattered data has been preserved and it is visually pleasing as well.

## 5   Conclusion

This study exploits the rational function with free parameters to visualize the monotone scattered data. The rational cubic function [11] is used as side-vertex as well as boundary curve interpolant, resulting six rational functions in each triangular patch. These rational functions are stitched as $C^1$ rational function with 12 free parameters over each triangular by side-vertex interpolation scheme proposed by Nielson [9]. The domain is triangulated by Delaunay triangulation method. Delaunay triangulation method provides the optimal triangulation of given scattered data by avoiding thin and elongated triangles. The partial derivatives at the vertices of triangles are estimated by the local derivative estimation scheme [7].

The only trace of use rational function with free parameters to visualize the scattered data is found in [10], where the authors developed a hit and trail method for the selection of free parameters to visualize the positive shape of scattered data. On contrary, in this paper data dependent sufficient conditions are derived on free parameters to visualize the shape of monotone scattered data in a given direction. Hence the choice of free parameters is automated. Moreover, the choice of the derivative is left at the wish of the user. Thus the method is equally applicable to data with and without derivatives. The scheme developed in [2] is restricted to the Lipschitz continuous functions, whereas, the presented scheme of this paper is applicable to all monotone functions. The section of Numerical Examples comprises of two monotone functions which are borrowed from [2]. It is observed that monotone scattered data visualization scheme presented in this paper produces more pleasing results than [2] for the same functions.

The authors in [8] to preserve the shape of monotone scattered data, converted the given scattered data into a rectangular grid by drawing horizontal and vertical lines through data sites, whereas, in the presented scheme there is no need to convert the scattered data into rectangular grid. But, the presented scheme of the paper is equally applicable to regular data(each rectangle can partitioned into two triangles).

# References

1. Beaston, R.K., Ziegler, Z.: Monotonicity preserving surface interpolation. SIAM Journal of Numerical Analysis 22, 401–411 (1985)
2. Beliakov, G.: Monotonicity preserving approximation of multivariate scattered data. BIT Numerical Mathematics 45, 653–677 (2005)
3. Carlson, R.E., Fritsch, F.N.: Monotone piecewise bicubic interpolation. SIAM Journal of Numerical Analysis 22, 386–400 (1985)
4. Delgado, J., Peña, J.M.: Are rational Bézier surfaces monotonicity preserving? Computer Aided Geometric Design 24, 303–306 (2007)
5. Fang, T.P., Piegl, L.A.: Algorithms for Delaunay triangulation and convex-hull computation using a sparse matrix. Computer Aided Design 24, 425–436 (1992)
6. Floater, M.S., Peña, J.M.: Monotonicity preservation on triangles. Mathematics of Computation 69(232), 1505–1519 (2000)
7. Goodman, T.N.T., Said, H.B., Chang, L.H.T.: Local derivative estimation for scattered data interpolation. Applied Mathematics and Computation 68, 41–50 (1995)
8. Han, L., Schumaker, L.L.: Fitting monotone surfaces to scattered data using $C^1$ piecewise cubics. SIAM Journal of Numerical Analysis 23(2), 569–585 (1997)

9. Nielson, G.: The side-vertex method for interpolation in triangles. Journal of Approximation Theory 25, 318–336 (1979)
10. Ong, B.H., Wong, H.C.: A $C^1$ Positivity Preserving Scattered Data Interpolation Scheme. In: Fontanella, F., Jetter, K., Laurent, P.J. (eds.) Advanced Topics in Multivariate Approximation, pp. 259–274. World Scientific Publishing Company, Singapore (1996)
11. Tian, M., Zhang, Y., Zhu, J., Duan, Q.: Convexity-preserving piecewise rational cubic interpolation. In: ISCIAS, Hefei, China (2005)

# Author Index

Asano, Tetsuo    103

Bessonov, N.    87

de Macedo Mourelle, Luiza    71

Gouko, Manabu    3

Hussain, Malik Zawwar    146, 156
Hussain, Maria    146, 156

Ito, Koji    3

Kannan, S.R.    127

Matsui, Kazuhiro    56
Montero, Elizabeth    41

Nedjah, Nadia    71
Neveu, Bertrand    41

Oliveira Costa Jr., Sergio    71

Pandiyarajan, R.    127

Ramathilagam, S.    127
Rao, Shrisha    114
Riff, María Cristina    41

Sarfraz, Muhammad    146
Sato, Haruo    56

Tanaka, Hiroshi    103

Veenhuis, Christian    20
Volpert, V.    87

GPSR Compliance

*The European Union's (EU) General Product Safety Regulation (GPSR) is a set of rules that requires consumer products to be safe and our obligations to ensure this.*

*If you have any concerns about our products, you can contact us on ProductSafety@springernature.com*

In case Publisher is established outside the EU, the EU authorized representative is:

Springer Nature Customer Service Center GmbH
Europaplatz 3
69115 Heidelberg, Germany

**Batch number: 09490872**

Printed by Printforce, the Netherlands